Developing Multi-Platform Apps with Visual Studio Code

Get up and running with VS Code by building multi-platform, cloud-native, and microservices-based apps

Ovais Mehboob Ahmed Khan

Khusro Habib

BIRMINGHAM—MUMBAI

Developing Multi-Platform Apps with Visual Studio Code

Commissioning Editor: Richa Tripathi
Acquisition Editor: Alok Dhuri
Senior Editor: Nitee Shetty
Content Development Editor: Tiksha Lad
Technical Editor: Pradeep Sahu
Copy Editor: Safis Editing
Project Coordinator: Deeksha Thakkar
Proofreader: Safis Editing
Indexer: Rekha Nair
Production Designer: Joshua Misquitta

First published: September 2020

Production reference: 1170920

Published by Packt Publishing Ltd.
Livery Place
35 Livery Street
Birmingham
B3 2PB, UK.

ISBN 978-1-83882-293-4

www.packt.com

`Packt.com`

Subscribe to our online digital library for full access to over 7,000 books and videos, as well as industry leading tools to help you plan your personal development and advance your career. For more information, please visit our website.

Why subscribe?

- Spend less time learning and more time coding with practical eBooks and Videos from over 4,000 industry professionals

- Improve your learning with Skill Plans built especially for you

- Get a free eBook or video every month

- Fully searchable for easy access to vital information

- Copy and paste, print, and bookmark content

Did you know that Packt offers eBook versions of every book published, with PDF and ePub files available? You can upgrade to the eBook version at `packt.com` and as a print book customer, you are entitled to a discount on the eBook copy. Get in touch with us at `customercare@packtpub.com` for more details.

At `www.packt.com`, you can also read a collection of free technical articles, sign up for a range of free newsletters, and receive exclusive discounts and offers on Packt books and eBooks.

Foreword

Of all the products I've had the privilege of working on at Microsoft in the last 25 years, Visual Studio Code has to be my favorite. The efficiency of the user interface combined with the power of the extensibility model creates a unique experience that lets developers decide what makes them productive. Whether you are new to programming or a seasoned professional developer, you will find time saving tools, tricks, and deep insights to make development fast and fun with Visual Studio Code.

Ovais and Khusro teach you the fundamentals of VS Code to quickly get you up and running and they show you how you can create and use extensions to VS Code to add new capabilities to make you more efficient. They show you how you can build and deploy real, modern day cloud-based applications with just an editor and they show you how you can elevate your game by using VS Code to develop on remote machines, operating systems, and even the cloud with GitHub Codespaces.

Chris Dias

Principal Group Program Manager, Visual Studio Code

Microsoft

Contributors

About the authors

Ovais Mehboob Ahmed Khan is a seasoned programmer and solution architect with nearly 20 years of experience in software development, consultancy and solution architecture. He has worked with various clients across the United States, Europe, Middle East and Africa. Currently he is working as a Sr. Customer Engineer at Microsoft, based in Dubai. He specializes mainly in Application development using .NET and other OSS technologies, Microsoft Azure and DevOps.

He is a prolific writer and has published few books on Enterprise Application Architecture, .NET Core and JavaScript, and written numerous technical articles on various sites. He likes to talk about technology and has delivered various technical sessions around the world.

> *I would like to thank my family for supporting and encouraging me in every goal of my life.*

Khusro Habib has been working in the IT industry for almost 20 years. He is a veteran programmer and delivered several enterprise grade solutions in the capacity of a consultant and solution architect. He is a certified Enterprise Architect Practitioner with the experience to design change for an enterprise.

His current focus is on Cloud Computing, Web and Mobile development, Data and Process Integration, Analytics and Machine Learning. With knowledge on various technology domains, he can place technology in the right context and address complex problems with workable IT solutions. He has a unique way of explaining things by breaking them down and then gradually building them up to create a better understanding.

> *I would like to dedicate this book to my family who have been very supportive throughout this journey.*

About the reviewer

Ankit Gupta has been involved in software development for over 8 years. During these years of experience, he has developed software solutions for low-code/no-code development platforms and virtual desktop infrastructure technology to simplify the user experience and customize systems.

As a senior developer, he reviews new feature specifications, user experience, performance, and accessibility, as well as helping his team with designs, code reviews, and following object-oriented programming concepts using the Agile model.

Packt is searching for authors like you

If you're interested in becoming an author for Packt, please visit `authors.packtpub.com` and apply today. We have worked with thousands of developers and tech professionals, just like you, to help them share their insight with the global tech community. You can make a general application, apply for a specific hot topic that we are recruiting an author for, or submit your own idea.

Table of Contents

Section 2:
Developing Microservices-Based Applications in Visual Studio Code

3

Building a
Multi-Platform Backend Using Visual Studio Code

4

Building a Service in .NET Core and Exploring Dapr

5

Building a
Web-Based Frontend Application with Angular

6

Debugging Techniques

7

Deploying Applications on Azure

8
Git and Azure DevOps

Section 3:
Advanced Topics on Visual Studio Code

9
Creating Custom Extensions in Visual Studio Code

10
Remote Development in Visual Studio Code

Preface

Developing Multi-Platform Apps with Visual Studio Code aims at exploring various functionalities of VS Code as an editor by taking the reader through a journey of developing a microservices based cloud native application.

The book is divided into three parts. The first part covers an introduction to VS Code, its features and the use of extensions. This creates a keen understanding of VS Code as an editor, by discussing the tooling features, increasing productivity and providing greater flexibility to a developer.

The second part covers building a cloud native application, based on multiple platforms. It starts by discussing the overall architecture and then moves on to developing the complete application. The application is based on a microservices architecture, where each backend service and the frontend is based on different technologies. The respective chapters elaborate the steps required to extend VS Code functionality to enable support for multiple platforms. The book also discusses Azure Event Hubs with Kafka protocol for inter-application messaging, Azure Kubernetes Services for container orchestration, Git and Azure DevOps for version management and deployment respectively. It also features Dapr for publishing messages to Kafka. Finally touching upon debugging the book explores the common VS Code debugging features along with the specific extensions required for each platform. All these topics are covered in this book in the most interactive manner to encourage you to try the examples yourself.

The last part covers advance topics related to VS Code such as creating custom extensions and remote development with examples.

Who this book is for

This book is for developers who are looking to learn how to effectively use Visual Studio Code to edit, debug, and deploy applications. Basic software development knowledge is a must to grasp the concepts covered in this book easily.

What this book covers

Chapter 1, Getting Started with Visual Studio Code, explains the key features of VS Code as an editor and how it differs from IDEs. The chapter covers VS Code in terms of the overall layout, command line options, multi-cursor editing, code refactoring, code navigation and user snippets among others.

Chapter 2, Extensions in Visual Studio Code. A key feature that makes Visual Studio Code more than just an editor is its extensibility framework. This allows the developer to add extensions relevant to their needs. This chapter will explain the extensibility framework, look at some important extensions, and explore how to add, delete, and configure extensions.

Chapter 3, Building a Multi-Platform Backend Using Visual Studio Code, starts off discussing the architecture and then building microservices on different platforms for the Job Ordering System we will develop in this part of the book. This will help you understand how the Visual Studio Code editor can be used in real-life scenarios.

Chapter 4, Building a Service in .NET Core and Exploring Dapr, develops a .NET Core hosted service to provide integration between different services. Communication is established between these services using a message broker system.

Chapter 5, Building a Web-Based Frontend Application with Angular, uses Visual Studio Code to develop a client-side web application and consumes the backend services developed in the previous chapters. The chapter will focus on how Visual Studio Code can be the editor of choice for frontend development.

Chapter 6, Debugging Techniques, focuses on exploring various debugging features of VS Code by setting up the environment for different technologies. The chapter also discusses different ways of launching the VS Code debugger.

Chapter 7, Deploying Applications on Azure, focuses on containerizing applications using Docker as the container technology and deploying to Azure Kubernetes Services for container orchestration scenarios.

Chapter 8, Git and Azure DevOps, focuses on the use of Git, understanding Git model and different strategies for source code version management. It also explores Azure DevOps for enabling CI/CD for automated build and deployment scenarios

Chapter 9, Creating Custom Extensions in Visual Studio Code, is about exploring the extension framework for extending VS Code functionality by developing custom extensions. The chapter takes you through the steps for building various types of extension such as using TypeScript, Code snippet and theme extension.

Chapter 10, Remote Development in Visual Studio Code, focuses on exploring the remote development feature of VS Code by developing applications on remote machines, containers and on cloud using GitHub Codespaces.

To get the most out of this book

The book deep dives into Visual Studio Code as an editor and explores its several features and functionalities. The objective of this book has been to learn tooling features by developing a real life application. The tool can be learned best with practice, and to get the most out of this book we encourage you to follow up each chapter by practicing the examples and also creating the application discussed in the book yourself. To help you along, the complete code base is available on the specified GitHub repository

The book uses the following software/hardwares:

Software/Hardware covered in the book	OS Requirements
Visual Studio Code	Windows, Mac OS, and Linux
Node.js	Windows, Mac OS, and Linux
Docker	Windows, Mac OS, and Linux
.Net Core 3.1	Windows, Mac OS, and Linux
JDK 1.8 or higher	Windows, Mac OS, and Linux
Java Maven 3.0 or higher	Windows, Mac OS, and Linux
Kubectl tooling	Windows, Mac OS, and Linux
Azure CLI	Windows, Mac OS, and Linux
Git	Windows, Mac OS, and Linux

If you are using the digital version of this book, we advise you to type the code yourself or access the code via the GitHub repository (link available in the next section). Doing so will help you avoid any potential errors related to the copying and pasting of code.

Download the example code files

You can download the example code files for this book from your account at www.packt.com. If you purchased this book elsewhere, you can visit www.packtpub.com/support and register to have the files emailed directly to you.

You can download the code files by following these steps:

1. Log in or register at www.packt.com.
2. Select the **Support** tab.

3. Click on **Code Downloads**.

4. Enter the name of the book in the **Search** box and follow the onscreen instructions.

Once the file is downloaded, please make sure that you unzip or extract the folder using the latest version of:

- WinRAR/7-Zip for Windows
- Zipeg/iZip/UnRarX for Mac
- 7-Zip/PeaZip for Linux

The code bundle for the book is also hosted on GitHub at `https://github.com/PacktPublishing/Developing-Multi-platform-Apps-with-Visual-Studio-Code`. In case there's an update to the code, it will be updated on the existing GitHub repository.

We also have other code bundles from our rich catalog of books and videos available at `https://github.com/PacktPublishing/`. Check them out!

Download the color images

We also provide a PDF file that has color images of the screenshots/diagrams used in this book. You can download it here:

`https://static.packt-cdn.com/downloads/9781838822934_ColorImagespdf.`

Conventions used

There are a number of text conventions used throughout this book.

`Code in text`: Indicates code words in text, database table names, folder names, filenames, file extensions, pathnames, dummy URLs, user input, and Twitter handles. Here is an example: "To push the complete history of changes to the remote repository, run `git push -u <remote_repo_name> <branch_name>`."

A block of code is set as follows:

```
- task: Docker@2
  displayName: Save Image
    inputs:
      command: save
      arguments: '-o $(Build.
```

```
ArtifactStagingDirectory)/$(NodeJSAPIName).
tar $(ContainerRegistryName)/$(NodeJSAPIName):$(Build.BuildId)'
```

Any command-line input or output is written as follows:

```
image deployment/$(k8jobreqdeployment)
$(k8jobreqdeployment)=$(ContainerRegistryName)/$(NodeJSAPIName)
:$(Build.BuildId)
```

Bold: Indicates a new term, an important word, or words that you see onscreen. For example, words in menus or dialog boxes appear in the text like this. Here is an example: "To get the credentials, click the **Generate Git Credentials** button in your repository on Azure DevOps. You will find this button on the **Clone Repository** page."

> **Tips or important notes**
> Appear like this.

Get in touch

Feedback from our readers is always welcome.

General feedback: If you have questions about any aspect of this book, mention the book title in the subject of your message and email us at customercare@packtpub.com.

Errata: Although we have taken every care to ensure the accuracy of our content, mistakes do happen. If you have found a mistake in this book, we would be grateful if you would report this to us. Please visit www.packtpub.com/support/errata, selecting your book, clicking on the Errata Submission Form link, and entering the details.

Piracy: If you come across any illegal copies of our works in any form on the Internet, we would be grateful if you would provide us with the location address or website name. Please contact us at copyright@packt.com with a link to the material.

If you are interested in becoming an author: If there is a topic that you have expertise in and you are interested in either writing or contributing to a book, please visit authors.packtpub.com.

Reviews

Please leave a review. Once you have read and used this book, why not leave a review on the site that you purchased it from? Potential readers can then see and use your unbiased opinion to make purchase decisions, we at Packt can understand what you think about our products, and our authors can see your feedback on their book. Thank you!

For more information about Packt, please visit `packt.com`.

Section 1: Introduction to Visual Studio Code

This section focuses on the basics of Visual Studio Code, starting with a brief introduction to Integrated Development Environments (IDEs) and editors, followed with some command-line options, tips, and tricks. It then covers key features of Visual Studio Code that will help learners develop complete knowledge about the tool.

This section comprises the following chapters:

- *Chapter 1, Getting Started with Visual Studio Code*
- *Chapter 2, Extensions in Visual Studio Code*

1
Getting Started with Visual Studio Code

Visual Studio Code or, as it is mostly called, **VS Code**, is one of the most popular coding tools today. It focuses on being fast and extendible, and can build up to the needs of a wider developer community.

Earlier, the focus was on providing developers with an extensive development environment that could integrate the complete software development life cycle, from writing code to deploying the solution. To increase productivity, several in-built features were provided to automate repetitive tasks. Most of the environments supported a particular language and an abundance of prebuilt features for that language.

As the industry moved from thick installable clients to web-based applications, the choice of development platforms and tools also changed. These browser-based web applications that were developed using **HyperText Markup Language (HTML)**, **Cascading Style Sheets (CSS)**, and JavaScript did not require heavy integrated development environments, but rather, simple and plain editors.

Looking at this change, Microsoft released its first cross-platform and multi-language support editor in 2015, called VS Code.

VS Code is a fast and lightweight editor that follows the concept of *take what you need*. The extension framework provides flexibility and has created a marketplace where extensive tooling features are available for use. It also allows the automation of build processes and has strong integration with Git for version control management. It allows developers to install and build an environment according to their specific needs.

In this chapter, we will take you through the basics of an editor and walk through the key difference between an editor and an **Integrated Development Environment (IDE)**. To get acquainted with VS Code, we will start by setting up the environment and exploring several options. Finally, we will highlight some tips and tricks that can increase your productivity. At the end of this chapter, you will be up and running with VS Code, have a good knowledge of how to configure the tool, and will be aware of some neat tricks. So, let's get started.

The main topics covered in this chapter are the following:

- Discussing editors and IDEs
- Discussing basic features of VS Code
- Setting up VS Code
- Exploring the VS Code layout
- Editing and code navigation in VS Code

Discussing editors and IDEs

Our focus in is this book is to explore and learn VS Code, but before we do that, let's discuss the difference between an editor and IDE and try to understand where and how VS Code is positioned for developers.

Over time, several languages and frameworks have become available for developers to work with. Along with these languages, a considerable amount of effort has been put in place to create the right tool to increase developer productivity and support the complete development life cycle.

These tools can be categorized into editors and IDEs.

Editors support a variety of languages, work around files and folders, and they are limited in terms of projects or solutions. They are lightweight and predominantly keyboard-centric, which allows developers to work faster.

IDEs support code editing, compiling, and debugging, as well as code execution. They are mainly specific to a particular language or a few selective languages. They usually work with project or solution files, provide support while writing code in terms of **IntelliSense**, and build processes that are well integrated into the environment. They provide language-specific wizards to help in generating the project skeleton and code while providing support for **Application Life Cycle Management** (**ALM**). In summary, IDEs provide extensive productivity features but are limited to a set of languages and frameworks.

VS Code falls between the editor and IDE space. It's lightweight and fast, yet provides support for several languages through a set of easily installable VS Code extensions. It's a cross-platform editor supporting Windows, macOS, and Linux operating systems, and provides features to cover the complete development life cycle.

Discussing basic features of VS Code

With a brief introduction to development tools, we will now start off our learning journey by exploring some basic features of VS Code.

Files and folders

Unlike an IDE, VS Code does not depend on creating a solution or a project file. Your project is your folder, with subfolders and files in it. You can use the **File** menu and **Open File...** or **Open Folder** to start editing, as illustrated in the following screenshot:

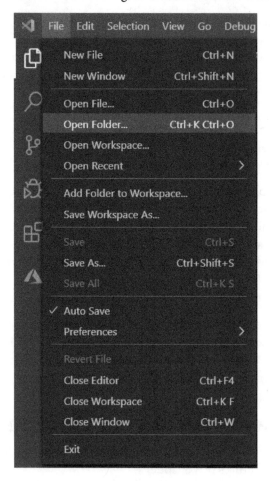

Figure 1.1 – File menu in VS Code editor

Opening the folder will load all the files of the folder in VS Code. You can open multiple folders at a time. These folders will be available in the left pane of the editor.

Workspace

Often, your product will be organized into multiple interlinked projects. In this case, VS Code provides an option to combine these folders in a **workspace**. This way, when you open a workspace, it will open all the related project folders.

To do this, you can simply select the **Add Folder to Workspace…** option and then use **Save Workspace As…** to save your workspace. This will generate a file of type .code-workspace that will contain all the folder names included in this workspace, as illustrated in the following screenshot:

Figure 1.2 – Project folders and the workspace file

Opening the workspace in the editor will open the linked project folders. The following screenshot shows how the workspace and folders are linked together in VS Code:

Figure 1.3 – Project folders part of workspace in the editor

The following code snippet shows the content of the .code-workspace file, for the preceding two projects:

```
{
    ''folders'': [
        {
            ''path'': ''Project 1''
        },
        {
            ''path'': ''Project 2''
        }
    ],
    ''settings'': {}
}
```

The .code-workspace file stores the path of folders linked in the workspace.

IntelliSense

One of the most powerful features for any developer is code completion: it comes in very handy when the development environment prompts the methods or attributes of a class, or suggests a list of keywords matching with what you are typing.

VS Code provides out-of-the-box IntelliSense for *JavaScript*, *TypeScript*, **Javascript Object Notation** (**JSON**), *HTML*, *CSS*, and **Sassy CSS** (**SCSS**). If this list does not satisfy your requirements, you can add a language extension of your choice from the marketplace. Just as with any other IDE, use *Ctrl + spacebar* to trigger the IntelliSense for the context.

Tasks

Compared to editors, IDEs provide automated and well-integrated processes that increase the productivity of a developer. These can include build processes, linting, and deployment, among others. VS Code provides similar functionality through its **Tasks** framework, which is illustrated in the following screenshot:

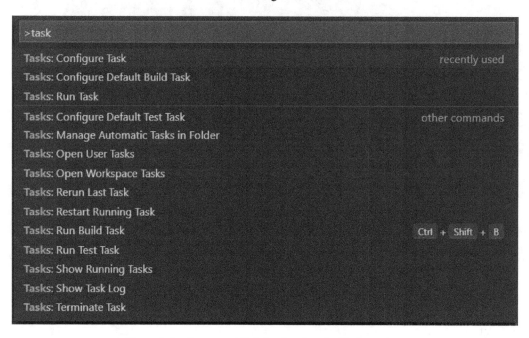

Figure 1.4 – Command Palette showing the Tasks option

Based on the files in your projects, it can detect tasks for build tools such as gulp, grunt, jake, and npm and also provides an option to create your own task runners using predefined templates. You can also link your tasks to the overall build process, which is triggered by *Ctrl + Shift + B*. Have a look at the following screenshot:

Figure 1.5 – Select a task template

The preceding screenshot shows the list of task templates to choose from.

Debugging

Another important feature of VS Code is the ability to debug code. Using VS Code, you can debug Node.js, JavaScript, and TypeScript out of the box, as illustrated in the following screenshot:

Figure 1.6 – Multiplatform debugging support

Apart from the preceding code base, you can install debugger extensions from the marketplace within the tool, as illustrated in the following screenshot:

Figure 1.7 – Debugger extensions on marketplace

Chapter 6, Debugging Techniques, is dedicated to exploring debugging in detail with VS Code.

Version control

VS Code integrates Git for version control. You can quickly initialize your Git repository by calling the command palette with *Ctrl + Shift + P* or *Command + Shift + P* and selecting **Git: Initialize Repository**, as illustrated in the following screenshot:

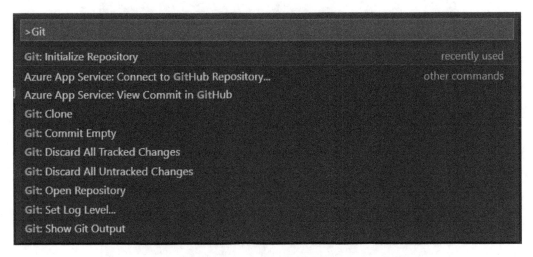

Figure 1.8 – Initialize Git repository using the command palette

To clone a repository, call the command palette and select **Git: Clone**, VS Code will request you to enter the **Uniform Resource Locator** (**URL**). After entering the URL, you can specify the folder or directory where you want your code to be cloned. The **Git: Clone** functionality can be seen in the following screenshot:

Figure 1.9 – Clone Git repository using the command palette

Additionally, you can also search for your favorite source code management tool extensions, as illustrated in the following screenshot:

Figure 1.10 – Integrate other source code management extensions from marketplace

The preceding screenshot shows a list of source code management extensions available in VS Code.

Keyboard shortcuts

Keyboard shortcuts provide quick accessibility of features; using them, you can navigate to options using commands instead of menus. To view a complete list of keyboard shortcuts, you can go to **Files | Preferences | Keyboard Shortcuts**. You can also use the pencil icon to change the shortcut.

We have now explored some basic features of VS Code. Next, let's move toward setting up our VS Code environment.

Setting up VS Code

VS Code is a cross-platform editor and supports Windows, macOS, and Linux OS. You can start by visiting `https://code.visualstudio.com`.

Based on the **operating system (OS)** you are using, the website will prompt you to download for your OS. If you would like to download for another OS, you can click on other platforms or press the down arrow button that shows next to the OS name, as illustrated in the following screenshot:

Figure 1.11 – Different versions of VS Code for different OSes

There are two versions available for download. The **stable** version is updated every month with new features, whereas the **insiders** version is a nightly build that provides an early peak into the upcoming features. Both versions can run side by side on the same machine.

Launching VS Code

By now, you should have downloaded your copy of VS Code and followed the guided installation wizard to set up your environment. Next, let's start off by looking into different ways of launching VS Code and explore some command-line options.

The simplest way of starting VS Code is by running the `code .` command.

This will open up a new instance of VS Code. If this command does not work in your macOS installation, you can follow the next steps. For Linux, you can visit `https://code.visualstudio.com` and look for **Setup | Linux**.

Setting up the command line for macOS

If you already have a `Bash profile`, please skip *Steps 1* and *2*. Otherwise, proceed as follows:

1. Write the `cd ~/` command to go to your home folder.
2. Write the `touch .bash_profile` command to create a new file.
3. Then, on the terminal window, write the following commands:

```
cat << EOF >> ~/.bash_profile
#Add Visual Studio Code (code)
Export PATH=''\$PATH:/Applications/Visual Studio Code.
app/Contents/Resources/app/bin
EOF
```

4. Close the terminal window and reopen to check whether the `code .` command works.

Now that your command line is set up and working, let's look at some different ways of launching VS Code.

The following command launches the code and opens the directory in VS Code where this is run:

```
code .
```

The `-r` variant allows you to open the specified workspace file in an already loaded VS Code instance; you can replace the workspace file with any file you would like to edit, as illustrated in the following code snippet:

```
code -r ''c:\My VS Code Projects\my project workspace.code-workspace''
```

The −n addition allows you to open a new instance of VS Code, as illustrated in the following code snippet:

```
code -n ''c:\My VS Code Projects\Project 1''
```

If you would like to open a particular file and go to a specific row and column, use the −g addition. The following command will launch a new instance of VS Code; open the launch.json file and place the cursor on row 5 and column 10:

```
code -n -g ''c:\My VS Code Projects\Project 1\.vscode\launch.
json'':5:10Exploring VS Code Layout
```

In this section, we will explore the editor's layout and the different panes. To get you familiar with the editor, we will go through the different sections of the editor and explain their utility. The layout of the editor can be seen in the following screenshot:

Figure 1.12 – VS Code editor layout

The most prominent section of the editor is the big pane on the right. This is the place where you edit code. Files selected from the EXPLORER are opened in this pane. You can open multiple files for editing at the same time.

Activity Bar

The toolbar on the left is called the Activity Bar. The Activity Bar provides quick access to important features of the tool.

The first button, 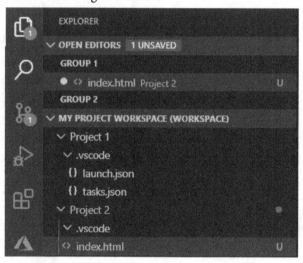, is the explorer, where you see the folders and files loaded in VS Code, as illustrated in the following screenshot:

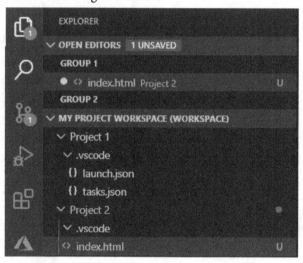

Figure 1.13 – Explorer

The number on the button shows the number of files changed and not saved. If **Auto Save** is enabled from **File** | **Auto Save** then this will not be shown, as every time a change is made, VS Code automatically saves it.

The search button allows you to search in files and folders. It searches the text inside a file. As you type, VS Code will filter the list of files containing the entered text, as illustrated in the following screenshot:

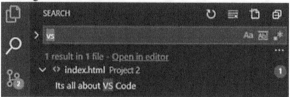

Figure 1.14 – Search text in files and folders

The next button is for version control. Git is very well integrated in the tool and provides some handy features.

If you have already initialized a Git repository or cloned an existing one, the source control button shows the changes made. We will be discussing Git in more detail in *Chapter 8, Git and Azure DevOps*. The button is illustrated in the following screenshot:

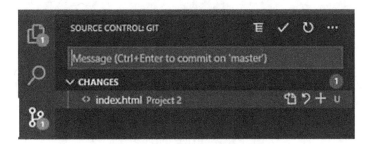

Figure 1.15 – Changes made to Git

To run and debug your code, use the debug button , which allows you to run the program and provides the usual debugging features. This is illustrated in the following screenshot:

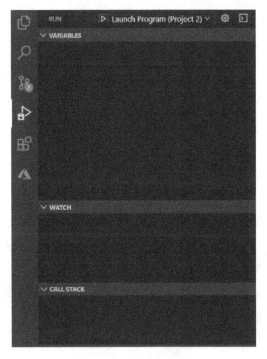

Figure 1.16 – Debugger

To manage your extensions in VS Code, you can use the extension button , which is illustrated in the following screenshot:

Figure 1.17 – Extensions pane

In *Chapter 2*, *Extensions in Visual Studio Code,* we will explore extensions in more detail.

Status Bar

At the bottom is the Status Bar. It contains the Git branch you are working on and errors and warnings, with an option to directly drill down to them. On the right, VS Code will show the type of file you are working with, a feedback option, and a notification button.

Quick Links

On the top-right corner (*Figure 1.12*), you have an option to see the code changes by pressing the **Open Changes** button, and you will be able to edit two objects side by side by pressing the **Split Editor** button.

Panel

The panel displays errors and warnings and debug-related information, and integrates a terminal window, as illustrated in the following screenshot:

Figure 1.18 – VS Code panel

Integrated terminal

VS Code integrates a terminal window inside the editor. This is a very handy feature as you don't need to open a terminal window separately to run command-line instructions. Most of the frameworks and platforms provide command-line tooling, and this feature really comes in handy. One such example is while creating your Angular frontend using the Angular **command-line interface** (**CLI**).

VS Code provides an option to create several terminal windows for the same project. The shortcut key for creating a new terminal window is *Ctrl + `*; for adding a new terminal window, the shortcut key is *Ctrl + Shift + `*.

You can also select your default terminal window, as illustrated in the following screenshot:

Figure 1.19 – Option to select the default shell

Once you press the **Select Default Shell** option from the menu, VS Code will show a menu to change your default shell, as illustrated in the following screenshot:

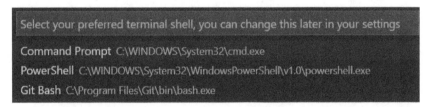

Figure 1.20 – Command Palette with a list of options for the terminal shell

Similarly, by pressing the ➕ button in *Figure 1.19*, you can create a new terminal window. VS Code will list the open project folders in the command palette to let you select which project you would like to create in the terminal window.

With *Ctrl/Command + Shift + 5*, you can split the bottom pane to show two terminal windows side by side. The same can be achieved by pressing the ⊞ button in *Figure 1.19*. It is possible to open several terminal windows side by side. To switch focus between the terminals, use *Alt/Option + Right Arrow Key* or *Alt/Option + Left Arrow Key*.

Split Editor

We have talked about the option to edit two documents side by side. VS Code takes this feature to the next level and allows a developer to open several windows side by side, as follows:

- *Ctrl/Command + 2*: This will open up the second editor.

- *Ctrl/Command + 3*: This opens up the third editor, all side by side.

You can go up to *Ctrl/Command + 8*. Each editor window is identified by the same keystroke, *Ctrl/Command + <number>*, which will allow you to switch between the editors.

Use *Ctrl/Command + W* to close the editor window; the one currently selected is closed.

Command palette

The command palette is one of the most important sections of the editor. VS Code is fast and focuses on keyboard-centric navigation; the command palette is the center piece of this editor.

You can quickly call the command palette by pressing *Ctrl + Shift + P* on Windows and *Command + Shift + P* on macOS.

On pressing this command, you will see a window pop up at the top of the editor initialized with a > sign, as illustrated in the following screenshot:

Figure 1.21 – Quick way to call a VS Code function using the command palette

This is a quick way of calling most of the editor options.

Search option: After you have opened the command palette, remove the > sign. Then, you will be able to search for files in the project. This is illustrated in the following screenshot:

Figure 1.22 – Search for files using the command palette

Help (?): After opening the command palette, type the question mark sign: you will see a list of commands you can run. This is illustrated in the following screenshot:

Figure 1.23 – List of commands

Caret (>) sign: This is the same as *Ctrl/Command + Shift + P*. From there, you can run any command (for example, running a task), as illustrated in the following screenshot:

Figure 1.24 – Command palette with default caret sign

debug: Use the debug command to launch the debugger or add a debug configuration, as illustrated in the following screenshot:

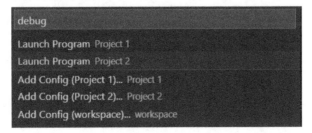

Figure 1.25 – Debugging options in the command palette

`ext`: This command helps you manage and install extensions, as illustrated in the following screenshot:

Figure 1.26 – Using the command palette to call extensions

We are almost done with the layout options in VS Code—let's explore themes before we close out this section.

Themes

Another important customization feature of the tool is the ability to set themes. Some developers like a dark colored environment, while others prefer light colors. VS Code provides you with a list of themes to choose from.

You can set your preferred theme by calling the command palette. Press *Ctrl/Command + Shift + P* to call the command palette, type `theme`, and then select **Preferences: Color Theme**, as illustrated in the following screenshot:

Figure 1.27 – Calling theme settings from the command palette

Here is a sample screenshot of the list of themes you will get to choose from:

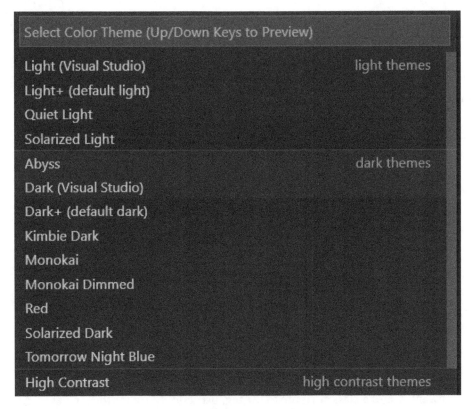

Figure 1.28 – List of themes to choose from in VS Code

We are now familiar with the editor's layout. Next, let's look at some important editing features of VS Code.

Basic editing in VS Code

VS Code comes with some great options, which will enable you to code faster. Let's look at some of the options.

Generating HTML

The Emmet 2.0 extension is built into the editor, and helps you to quickly write HTML code.

Here are some examples.

To create a table with five rows having two columns in each row, you can use the following statement:

```
table>tr*5>td*3
```

This is what the table looks like:

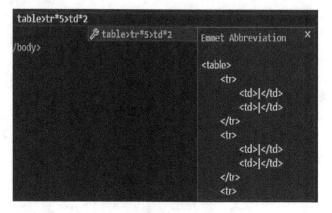

Figure 1.29 – Generate HTML using the Emmet extension

As shown in the preceding screenshot, you will notice that VS Code starts showing you the HTML it will generate. Press *Enter* to generate the HTML.

In case you want to add a class attribute to the tag, use a dot (.) after the tag name. An example of this can be seen in the following code snippet:

```
table>tr*5.myclass>td*2
```

This will generate the same table as in *Figure 1.29*, with myclass placed as a class attribute for the <tr/> tag. The cursor will be placed between the first <td></td> tag; use *Tab* to navigate to the next <td/> tags for faster editing.

Multi-cursor editing

One of the most important features of VS Code is the multi-select cursor. It has several variants, which we will explore now, as follows:

- *Alt/Option + Ctrl/Command + Down Arrow Key*: This option will allow you to place a cursor on multiple lines at the same place. In the following screenshot, you can notice the cursor is placed on five different lines at the same spot:

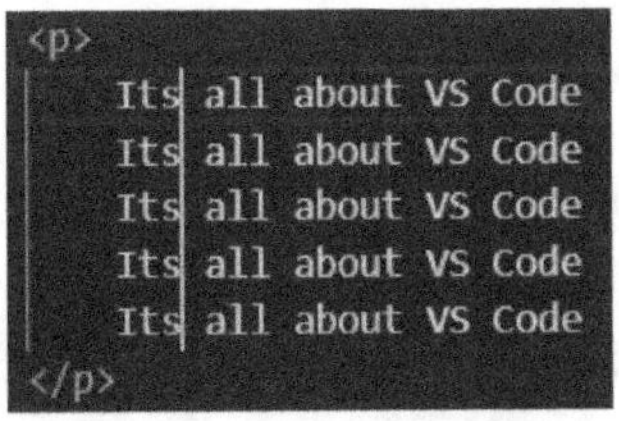

Figure 1.30 – Multi-cursor editing on the same position

- *Ctrl/Command + Shift + L*: This command will allow you to select multiple occurrences of the same text and edit at the same time. Pressing the arrow key will keep the cursors active and allow quick multi-cursor editing. Press *Esc* to remove the cursors. This is illustrated in the following screenshot:

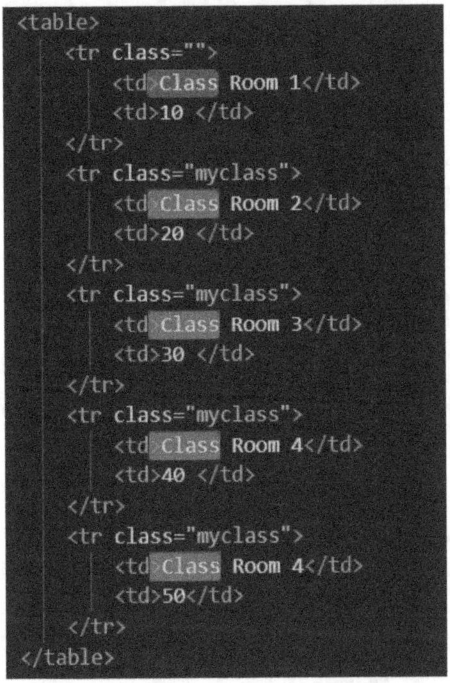

Figure 1.31 – Multi-cursor editing of the same text

- *Alt/Option* + mouse click: This option will allow you to place cursors at specific points and edit at the same time. Press *Esc* to remove the cursor. This is illustrated in the following screenshot:

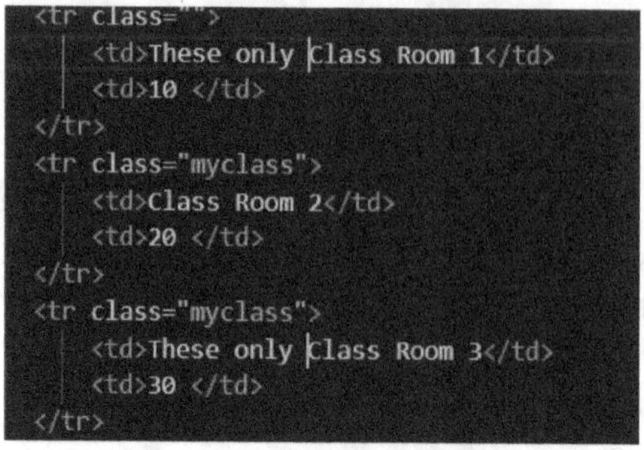

```
<tr class="">
    <td>These only Class Room 1</td>
    <td>10 </td>
</tr>
<tr class="myclass">
    <td>Class Room 2</td>
    <td>20 </td>
</tr>
<tr class="myclass">
    <td>These only Class Room 3</td>
    <td>30 </td>
</tr>
```

Figure 1.32 – Multi-cursor editing by placing cursor on mouse clicks

- *Ctrl/Command + D*: Highlight a text item and then, on each press of *Ctrl/ Command+D*, the system will move the cursor to select the same text item. If you would like to skip a specific selection, press *Ctrl/Command + K*, and then continue with *Ctrl/Command + D*.

- *Alt/Option + Shift + Down Arrow Key*: Place the cursor on a particular line and press the command to duplicate the same line.

- *Alt/Option + Down Arrow Key*: Place the cursor on a particular line and press the command to move the line below. The same works for multiple selected lines.

Code refactoring

While writing code, there are opportunities when the code written can be further optimized to increase readability and maintainability. Readable code is easier to change. It allows other developers to easily find the required code section and make changes as required. In a software development life cycle, having to often maintain or change code is more time consuming then first writing it.

VS Code provides some nice code refactoring features. Support for JavaScript and TypeScript is built into the tool. Let's go through a few examples.

Extracting to a constant

Once you select a section of code, a light bulb appears on the left; on clicking that, you will see a context menu open up with some refactoring options. Once you select the **Extract to constant in module scope** option, VS Code will create a constant and replace it with the string literals—for example, 'Product List!', shown as follows:

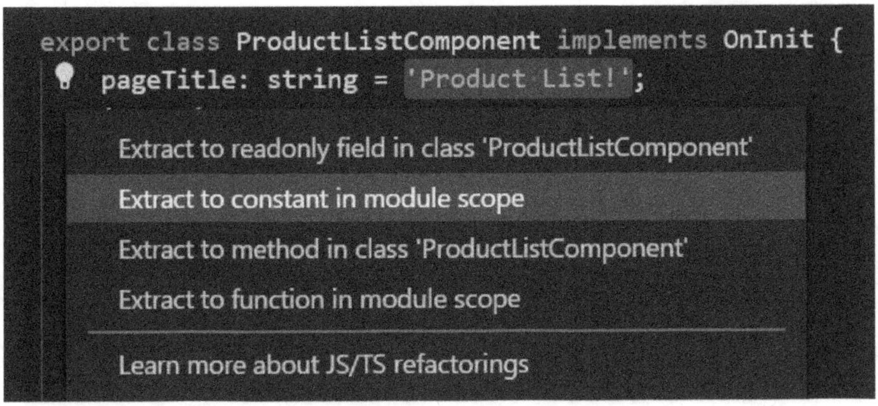

```
export class ProductListComponent implements OnInit {
💡    pageTitle: string = 'Product List!';

        Extract to readonly field in class 'ProductListComponent'

        Extract to constant in module scope

        Extract to method in class 'ProductListComponent'

        Extract to function in module scope

        Learn more about JS/TS refactorings
```

Figure 1.33 – Extract to constant

The following screenshot shows the refactored code. Here, string literals have been replaced with a constant variable:

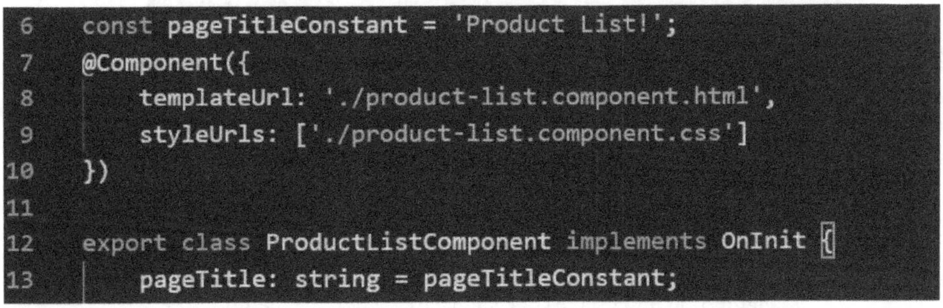

```
6    const pageTitleConstant = 'Product List!';
7    @Component({
8        templateUrl: './product-list.component.html',
9        styleUrls: ['./product-list.component.css']
10   })
11
12   export class ProductListComponent implements OnInit {
13        pageTitle: string = pageTitleConstant;
```

Figure 1.34 – Extract to constant refactored code

Apart from extract to constant, another option is extracting code to a method.

Extracting to a method

Most of the time while writing code, you come across a piece of code that can be reused. This code is extracted into a method to avoid code duplication. This enhances code maintainability, since future changes are not required to be done in several places.

VS Code provides a very easy-to-use and quick way to extract your code into reusable methods.

In the following example, the displayed success message is selected. A light bulb appears on the left and shows an **Extract to method in class** option:

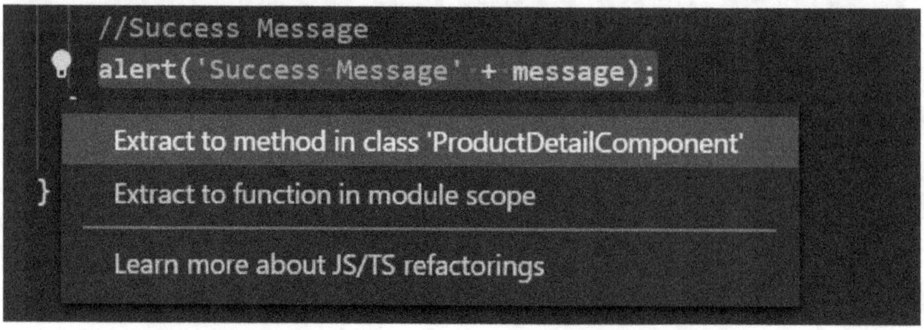

```
//Success Message
alert('Success Message' + message);

    Extract to method in class 'ProductDetailComponent'

    Extract to function in module scope

    Learn more about JS/TS refactorings
}
```

Figure 1.35 – Reusable code can be extracted to a method

Once you select this option, VS Code will ask for a method name as an input, as illustrated in the following screenshot:

```
//Success Message
this.showMessage(message);




private showMessage(message: string) {
    alert('Success Message' + message);
}
```

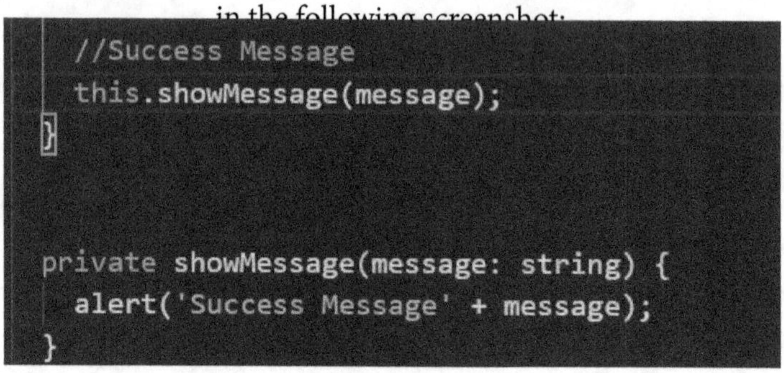

Figure 1.36 – Reusable code can be extracted to a method

On entering the method name, and as shown in preceding screenshot, the selected piece of code will be extracted into a method, and a method call will be placed in the same section.

Renaming symbols

It is often the case that while writing code, you might feel the need to rename some used variables. These variables are used in several places, and *Find and Replace* seems to be a hectic option.

VS Code here provides a quick **Rename Symbol** option.

Select the variable to be changed and press *F2*. VS Code will ask for the new variable, as illustrated in the following screenshot:

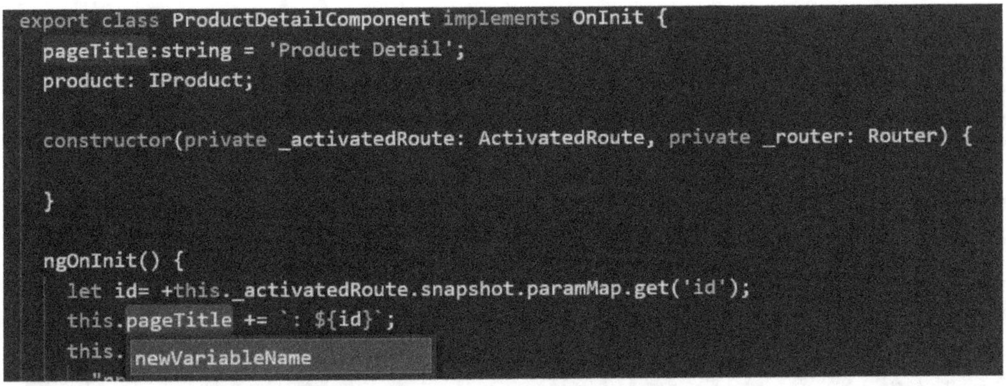

```
export class ProductDetailComponent implements OnInit {
    pageTitle:string = 'Product Detail';
    product: IProduct;

    constructor(private _activatedRoute: ActivatedRoute, private _router: Router) {

    }

    ngOnInit() {
        let id= +this._activatedRoute.snapshot.paramMap.get('id');
        this.pageTitle += `: ${id}`;
        this. newVariableName
```

Figure 1.37 – Rename variable using F2

On pressing *Enter*, the variable name is replaced throughout the file, as illustrated in the following screenshot:

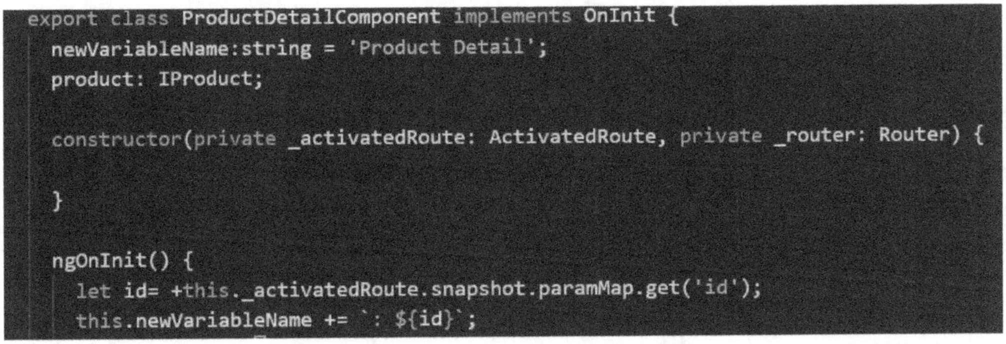

```
export class ProductDetailComponent implements OnInit {
    newVariableName:string = 'Product Detail';
    product: IProduct;

    constructor(private _activatedRoute: ActivatedRoute, private _router: Router) {

    }

    ngOnInit() {
        let id= +this._activatedRoute.snapshot.paramMap.get('id');
        this.newVariableName += `: ${id}`;
```

Figure 1.38 – Variable name changed to newVariableName

The preceding screenshot shows `pageTitle` changed to `newVariableName`.

Refactoring extensions

Furthermore, VS Code supports refactoring for other languages with extensions. You can search the refactoring extension for your language on the VS Code extension marketplace.

Snippets

Code snippets help you code faster—these are predefined code templates that VS Code suggests while writing code. The following is an example of a method template—on pressing *Tab*, the code on the right will be inserted:

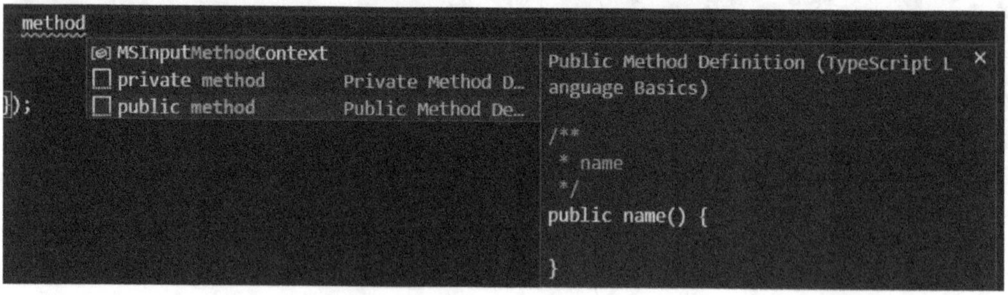

Figure 1.39 – Code snippets

Apart from the predefined templates, there are several extensions available that will come with their own code snippets.

Custom snippets

Snippets help you to be quick in writing repetitive code, and in case the out-of-the-box snippets do not satisfy your requirements, VS Code provides you with an option to define your own custom snippets, called **User Snippet**.

Let's create a user snippet for HTML code.

Go to **File | Preferences | User Snippet**. The command pallet will pop up. Enter html and press *Enter*, as illustrated in the following screenshot:

html

html.json (HTML)

Figure 1.40 – Create custom HTML code snippets

The html.json file contains the HTML user snippets. We will create a snippet that will generate a table. The list that follows details the different parts of a user snippet.

In our example, the following applies:

- `table-snippet`: This is the name of the snippet.
- `prefix`: The name mentioned in this section is used to call the snippet.
- `description`: This will be displayed in the pop-up window.
- body: This is the place where you write your code template. It's an array that will contain the line of code to be inserted when you call the snippet. \t is used for code indentation.

Copy and paste the following code into an `html.json` file:

```
''table-snippet'':{
    ''prefix'': ''mytable'',
    ''description'': ''Table Snippet'',
    ''body'': [
        ''<table>'',
        ''\t<tr>'',
        ''\t\t<td>Column 1</td>'',
        ''\t\t<td>Column 2</td>'',
        ''\t</tr>'',
        ''</table>'',
    ]
}
```

So, our snippet is created. Next, we will call it. If the following code does not work for you, restart VS Code.

As soon as you start typing `mytable`, VS Code suggests the code snippet you created, as illustrated in the following screenshot:

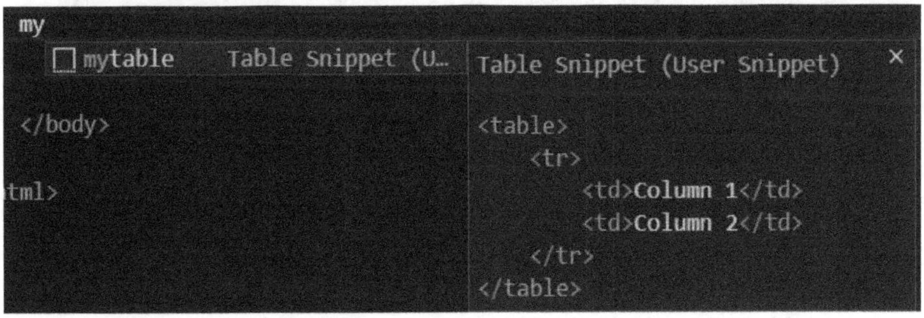

Figure 1.41 – Calling the custom HTML code snippet created earlier

On pressing *Tab*, the code will be inserted, as illustrated in the following screenshot:

```
<table>
    <tr>
        <td>Column 1</td>
        <td>Column 2</td>
    </tr>
</table>
```

Figure 1.42 – Code created from the custom HTML code snippet

The preceding screenshot shows the table generated from **User Snippet**.

Editing and code navigation in VS Code

While writing code, developers are often working with several files at the same time. Quickly navigating between files or within the same file is an important feature.

VS Code offers multiple options for code navigation—let's explore them one by one.

Go to line

To move the cursor to a particular line and column, use the Goto (:) Line command, as illustrated in the following screenshot:

```
<>  :6:15

Pr  Go to line 6 and column 15.

    2       <head>
    3           <title>Visual Studio Code</title>
    4       </head>
    5       <body>
    6       <h1>Hello and welcome to visual studio</h1>
    7       </body>
    8
    9
    10  </html>
```

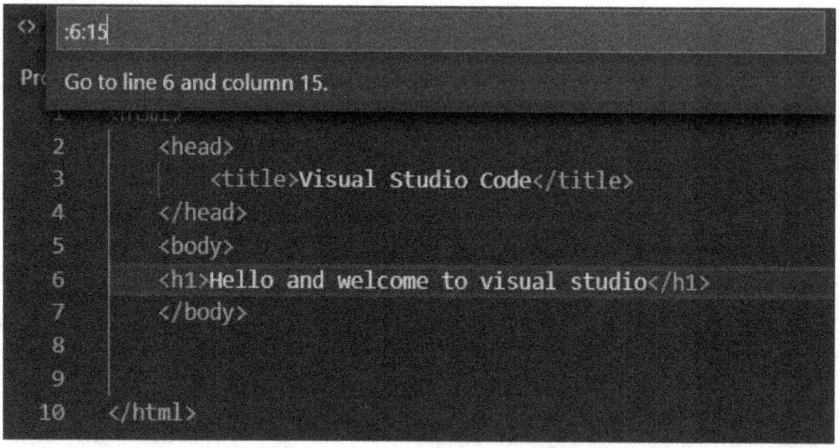

Figure 1.43 – Using the command palette to move cursor to a line and column

Go to symbol

To show the symbols used in your code and to navigate directly to a selected symbol, you can use **Go to Symbol (@) in a File**.

Also, while you are traversing, VS Code will highlight the code section, as illustrated in the following screenshot:

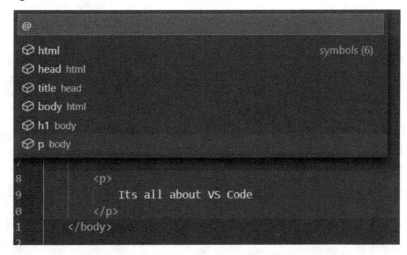

Figure 1.44 – Using the command palette to jump to a section of code

In the case of JavaScript, the command palette will show methods, as illustrated in the following screenshot:

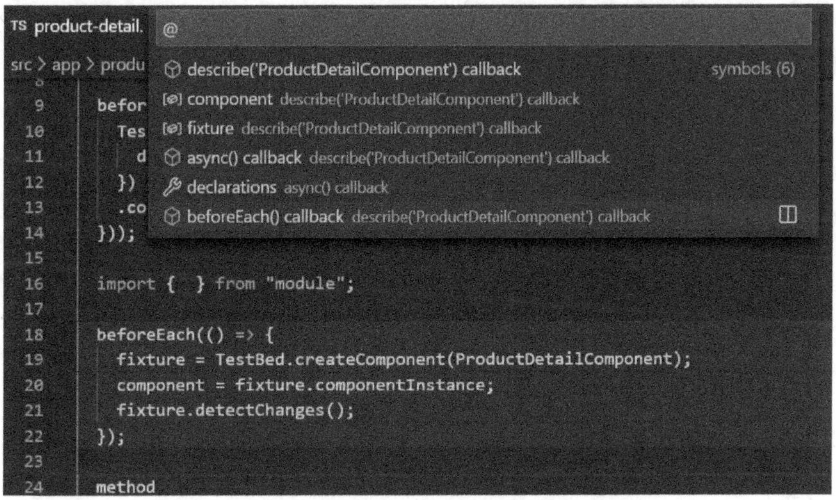

Figure 1.45 – Jump to code section in a JavaScript example

Using commands for navigation

Here are some useful commands for quick code navigation:

- *Ctrl/Command + F12*: Use this command to select an object and jump to its implementation.

- *Ctrl/Command + Shift + F12*: This is a quick way to peek into an implementation. The pop-up window also allows code editing. This is illustrated in the following screenshot:

```
 9       beforeEach(async(() => {
10           TestBed.configureTestingModule({
11             declarations: [ ProductDetailComponent ]
```
product-detail.component.ts C:\Users\10101\Desktop\AngularFirst\src\app\products - **Implementations (1)**
```
 2       import { IProduct } from './product';
 3       import { ActivatedRoute, Router } from '@angular/router';
 4
 5       @Component({
 6         templateUrl: './product-detail.component.html',
 7         styleUrls: ['./product-detail.component.css']
 8       })
 9       export class ProductDetailComponent implements OnInit {
10         pageTitle:string = 'Product Detail';
11         product: IProduct;
12
13         constructor(private _activatedRoute: ActivatedRoute, private _router: Router) {
14
15         }
16
```

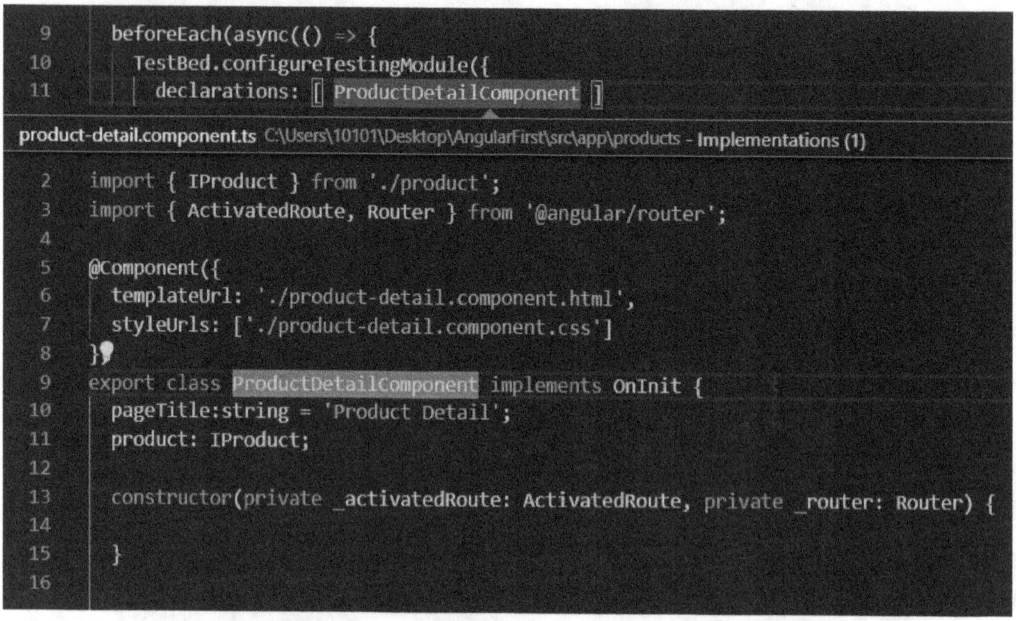

Figure 1.46 – Peek into an implementation

The preceding screenshot shows an example where the `ProductDetailComponent` implementation is open in **Quick Peek** mode.

Navigating between files

VS Code provides multiple options for navigating between different files. Let's assume you have a bunch of files open in your editor. To select a particular file from currently opened files, hold *Ctrl* and then press *Tab*. VS Code will open a list of files, and you can move your selection by holding *Ctrl* and pressing *Tab*. An example list of files can be seen in the following screenshot:

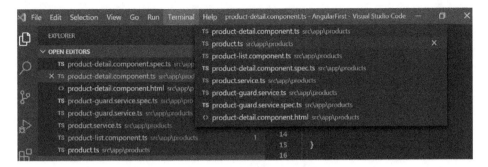

Figure 1.47 – Files listed for navigation

Apart from using *Ctrl* (hold) + *Tab*, you can also sequentially switch between multiple files opened in the editor.

Use *Alt/Option + Right Arrow* or *Alt/Option + Left Arrow* to switch from one file to other opened files.

Commenting code

Last but definitely not least, how to quickly comment and uncomment code by using the following keyboard shortcuts:

- *Ctrl/Command + K* then *C*: Use this command to comment a line.
- *Ctrl/Command + K* then *U*: Use this command to uncomment a line.

Summary

In this chapter, we started by looking at the editor and IDE space, and where VS Code falls. Then, we looked at the basic features of VS Code and moved on to setting up our environment. With VS Code being a multiplatform editor, we explained some steps related to macOS, and also provided guidance in case you are using a Linux OS. Next, we explored multiple ways of launching VS Code and also looked at several command-line options. Explaining the overall layout, we then moved on to showcasing some editing and code refactoring features. Finally, we closed out the chapter by looking at quick code navigation features.

By the end of this chapter, you should now have a basic background about development tools and a more detailed knowledge of VS Code as an editor. We learned about the basic features, layout, editing, and code navigation options.

Next, we will look in further detail at the VS Code extensions framework and how it can help you in enhancing your development experience.

2
Extensions in Visual Studio Code

Visual Studio Code (**VS Code**) is one of the most popular advanced editors in the industry and is used by many developers. One of the key features of VS Code is **extensions**. With extensions, any developer of any platform can use VS Code with any language and framework to develop any kind of application. You can install any extension from VS Code's extension gallery in your editor and use it to perform the desired operation.

There are various extensions available that are categorized under the labels Azure, Debuggers, Formatters, Keymaps, Language Packs, Programming Languages, Themes, and many more. You can add these extensions to enable extensibility in VS Code and use the features to make developer experience better. In addition, you can also write your own extensions for specific scenarios and share them in the marketplace so they can be used by other developers around the globe.

This extensibility and complaisance make VS Code one of the most powerful editors in the industry. In this chapter, we'll learn how to install and configure extensions and extension filters and will look at the various extensions that can be used to improve the productivity of developers.

Here are the following main topics that we will cover:

- Managing and configuring extensions
- Extension filters
- Client-side framework extensions
- Visual adjustment extensions
- Productivity extensions

Managing and configuring extensions

VS Code offers various types of extensions. Each extension is built for a specific purpose and requirement. In this topic, we will cover how you can browse, install, and manage extensions in your project. Once the VS Code is installed, you can open it by just typing code from the command prompt. You can then use the **CLI** (short for **command-line interface**) for different frameworks to create projects on the fly.

To open the **Extensions** bar, use the *Ctrl + Shift + X* command. This command opens the **Extensions** tab, as shown in the following screenshot, where you can search for any extension and add it to your editor:

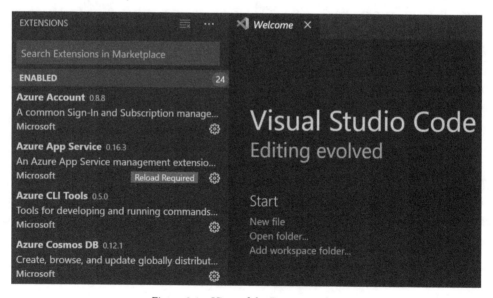

Figure 2.1 – View of the Extensions bar

This bar categorizes extensions as **ENABLED, RECOMMENDED,** and **DISABLED.** All of the extensions that are currently installed are placed under the **Enabled** tab. Some new recommended extensions will be listed under the **Recommended** tab, while the **Disabled** tab shows the extensions that are currently disabled. You can search for new extensions by entering the extension name. When you search for extensions, you can see brief information about the extension identifier, publisher, description, downloads, rating, and links to its repository and license.

> **Important note**
>
> There are many extensions that are available, published by many developers and enterprises. One tip when choosing extensions is to make sure that the review or rating is good, or that the extension is from a trusted publisher.

Suppose I want to write code in C/C++. I can search for the extension by typing C++. It will list all of the extensions named C++, and you can choose the one that is from a trusted publisher. In our case, it is from Microsoft:

Figure 2.2 – Adding a C/C++ extension

The preceding screenshot shows the C/C++ extension published by Microsoft with more than 99 million downloads and a rating of 5 stars (on a scale of 1–5) and the links to its actual repository where you can see more information about the extension and contribute to it. Finally, the license information helps you understand whether it is free to use commercially or you need a paid license for it.

Extension packs

Extension packs are a collection of extensions that enable developers to easily install the related extensions with a single installation. For example, the Azure extension pack can be used when working with Azure resources in VS Code. With one single extension pack, you can work with Azure App Services, functions, Docker tools, storage, data, Visual Studio team services, Azure Terraform tools, **IoT** (**internet of things**) tools, **AI** (**artificial Intelligence**) tools, and many more.

Installing extensions from VSIX

In the previous section, we saw how extensions can be installed from the VS Code extensions marketplace. However, there are scenarios where the extension is available offline and we need to add it to the VS Code project. In this case, we can manually install the extension through a command-line option or from the VS Code Command Palette dialog.

Installing extensions from the CLI

To install extensions from the command line, we use the following syntax:

```
code --install-extension {path of extension VSIX file}
```

Here is an example of installing a custom extension packaged in a `customextension.vsix` file:

```
code --install-extension customextension.vsix
```

In the preceding instances, the command `code` is the CLI of VS Code editor and `--install-extension` is the switch that can be used to install extensions manually.

Installing extensions from the Command Palette

To install extensions from the **Command Palette**, select (**…**) from the **Extensions** bar and click on the **Install from VSIX…** option, as shown in the following screenshot:

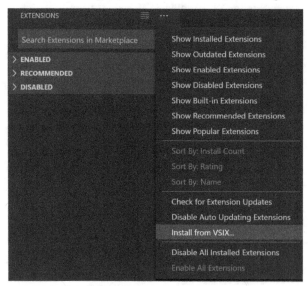

Figure 2.3 – Context menu showing different extension options

It will ask you to browse for the VSIX file from your machine and then install it.

Extension information

If the extension is not an extension pack, then you can see the complete details about the extension, such as the contributions and changelog; otherwise, you will get two tabs, namely **Details** and **Extension Pack**. Here are the details of each tab.

Details provides adequate information about the extension in terms of getting started with the extension. It provides information about what is needed to use the extension, quick links, FAQs, known issues, guidelines to contributing, the code of conduct, and other details:

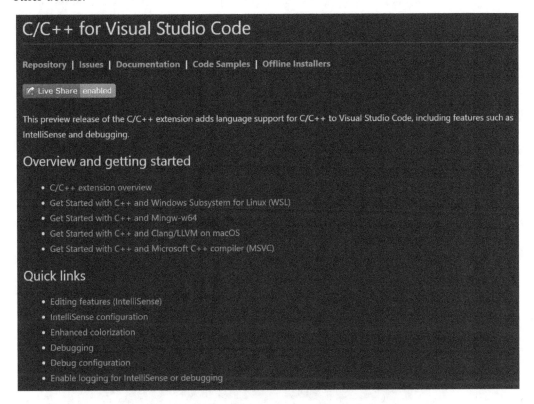

Some of the extensions also show a Live Share icon. Live Share is a feature that allows multiple developers to collaborate and contribute to the same file from their own editor. We will be discussing this later in this book.

The Contributions tab shows the extension commands, settings, keyboard shortcuts, debuggers, and views that are required to understand and use the extension correctly.

The Changelog provides the complete history about the changes that were made in terms of improvements, new additions, and bug fixes, and how the extension evolved with time.

The Extension Pack section contains all of the extensions that are installed with the pack.

Extension filters

In this topic, we will discuss some commands to filter extensions in VS Code.

View commands to filter extensions

VS Code offers various commands to filter extensions. You can open the **Commands** context menu by hitting the **...** from the **Extensions** bar that displays the list of commands to further filter extensions, as shown in the following screenshot:

Figure 2.5 – Context menu showing the different commands for extensions

Each option is self-explanatory and shows extensions based on different options. They allow you to check for extension updates, disable the autoupdating of extensions, install any custom extensions from a VSIX file, or disable all installed extensions.

Extension identifiers

VS Code also provides lot of identifiers that you can type in the **Search Extensions** textbox and find or filter the extension based on the identifier type. To list all of the identifiers, you can type @ and it will give you options to select from and filter extensions accordingly:

Figure 2.6 – Using extension identifiers

Here are some of the most commonly used ones:

- @id: If you know the ID of the extension, then you can use the @ id:{extension_code} to add the C/C++ extension type, such as @id: ms-vscode.cpptools.

- @builtin: Shows out-of-the-box extensions that come with VS Code.

- @sort: Sorts extensions based on their name, ratings, and installs. For example, to sort C++ extensions based on their name, you can type C++ @sort:name. You can also sort them based on their rating or installs by running the C++ @ sort:rating and C++ @sort:installs commands respectively.

- @recommended: Shows recommended extensions.

- @enabled: Shows a list of enabled extensions.

- @disabled: Shows a list of disabled extensions.

- @installed: Shows a list of installed extensions.

- `@tag`: Search for extensions with tags. For example, `@tag:Microsoft` will search for all of the extensions that are tagged with Microsoft.

- `@outdated`: Shows a list of outdated extensions.

You can also combine multiple extensions to filter or refine your search for more accurate searching. For example, to search enabled extensions with category formatters, we can type `@enabled @category:formatters`.

Filtering extensions using categories

Extensions can also be filtered with categories and tags. Categories and tags are used to describe the features of the extension and then you can use the given syntax to filter out extensions based on the categories or tags assigned to them.

To filter extensions based on categories, you can type `@category:` in the extension bar and filter it out based on the supported categories shown in the following screenshot:

Figure 2.7 – Using the category filter

Similarly, you can also use `@tag:` to filter out extensions based on their tag. For example, you can filter out the extensions tagged to Microsoft, as shown in the following screenshot:

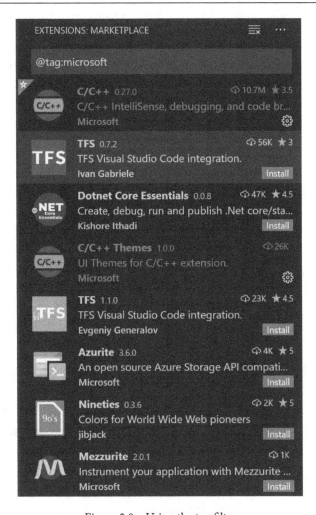

Figure 2.8 – Using the tag filter

Unlike categories, when searching with @tag:, we won't get any IntelliSense extensions, and you need to search the marketplace in order to know which tags are helpful.

Creating a list for recommended extensions

When working in a team, we sometimes need to share a list of recommended extensions with the other members of the team. There might also be the case where a lead developer adds certain extensions that are essential for a project and wants to share those with the team. One option is to compile the list of extensions and distribute it to the team so they can install them on their own machines. Another way of doing this is to create an extension list and share it with your team.

To create a list of recommended extensions, open the Command Palette by hitting *F1* and then type `Extensions: Configure Recommended Extensions (Workspace)`. This command creates a new `extensions.json` file that will reside in the `.vscode` folder. You can open this file and add the recommended extensions by adding the following snippet:

```
{
    "recommendations": [
        "ms-kubernetes-tools.vscode-kubernetes-tools",
        "azuredevspaces.azds"
    ]
}
```

Here is the list of default paths in which the `.vscode` folder resides for each platform:

- **Windows**: `%USERPROFILE%\.vscode\extensions`
- **Linux**: `~/.vscode/extensions`
- **macOS**: `~/.vscode/extensions`

The default location for extensions can be changed by running the following command:

```
code --extensions-dir <dir>
```

The following example sets the default extensions directory to the `vscodeextensions` folder in the C drive:

```
code --extensions-dir c:\vscodeextensions
```

Each extension should be added by providing its extensions code. You can then share this list with any member in your team who can place it under the `.vscode` folder. Finally, once the VS Code opens, it will show users the recommended extensions, which can be installed right away:

Figure 2.9 – List of recommended extensions

This is one of the best features when we are working as a team. Every team of developers can build their own set of recommended lists of extensions and share with other team members so that they can use it.

Client-side framework extensions

VS Code is very popular for open source development. Being an advanced editor, it provides many features that are usually available only in full-fledged IDEs. One of its core features is IntelliSense, which can be added via necessary extensions for specific platforms. In this section, we will discuss some code snippet extensions that can be used alongside different client-side frameworks when building applications on top of it. Whether you are working on plain vanilla JavaScript-based projects or creating an application on Angular or React Native, these extensions are useful and can help you generate code snippets on the fly for faster development.

Code snippet extensions for client-side frameworks

There are several extensions that are available for various frameworks and platforms. With snippet extensions, we can generate code on the fly by just typing or using their short code. For example, with the JavaScript (ES6) code snippet extension in the StandardJS Style extension, we can easily generate a snippet to generate a `forEach` loop in ES6 syntax.

> **Tip**
> Write some code and hit *Tab* to generate the snippet.

Some of the most common code snippet extensions that are used for building projects on JavaScript, Angular and React Native are discussed in the following sections.

JavaScript code snippets

This extension contains code snippets for JavaScript in ES6 syntax for the VS Code editor. Some of the snippets and their triggers are shown in the following list:

- `cas`: Console alert method – `console.assert(expression, object)`
- `ccl`: Console clear – `console.clear()`
- `cco`: Console count – `console.count(label)`
- `cdi`: Console `dir console.dir`
- `cer`: Console error – `console.error(object)`
- `cgr`: Console group – `console.group(label)`
- `cge`: Console `groupEnd` – `console.groupEnd()`
- `clg`: Console log – `console.log(object)`
- `clo`: Console log object with the name – `console.log('object :', object);`
- `ctr`: Console trace – `console.trace(object)`
- `cwa`: Console warn – `console.warn`
- `cin`: Console info – `console.info`
- `clt`: Console table – `console.table`
- `cti`: Console time – `console.time`
- `cte`: Console `timeEnd` – `console.timeEnd`

You can refer to `https://github.com/xabikos/vscode-javascript` to see the complete list of JavaScript code snippets.

React Native code snippets

For React Native projects, the ES7 React/Redux/GraphQL/React Native snippet is one of the best extensions that provides collections of React Native code snippets, which are useful for React Native/React/Redux ES6/ES7 and flowtype/TypeScript, Storybook.

Some of the snippets and their methods are shown in the following list:

- `imr` - `import React from 'react'`
- `imrd` - `import ReactDOM from 'react-dom'`
- `imrc` - `import React, { Component } from 'react'`
- `imrcp` - `import React, { Component } from 'react'` & `import PropTypes from 'prop-types'`
- `imrpc` - `import React, { PureComponent } from 'react'`
- `imrpcp` - `import React, { PureComponent } from 'react'` & `import PropTypes from 'prop-types'`
- `imrm` - `import React, { memo } from 'react'`
- `imrmp` - `import React, { memo } from 'react'` & `import PropTypes from 'prop-types'`
- `impt` - `import PropTypes from 'prop-types'`
- `imrr` - `import { BrowserRouter as Router, Route, NavLink} from 'react-router-dom'`
- `imbr` - `import { BrowserRouter as Router} from 'react-router-dom'`
- `imbrc` - `import { Route, Switch, NavLink, Link } from react-router-dom'`
- `imbrr` - `import { Route } from 'react-router-dom'`

You can refer to `https://github.com/dsznajder/vscode-es7-javascript-react-snippets` to see the complete list of React Native code snippets.

Angular code snippets

For Angular code snippets, you can check out Angular Snippets (version 9), developed by John Papa. This is one of the most popular extensions that is used with Angular projects. This extension contains collection of snippets that are used to generate code for Angular projects.

Here is a list of some of the snippets and their purpose that you can use them for in your projects:

- `a-component`: Component
- `a-component-inline`: Component with inline template
- `a-component-root`: Root app component
- `a-ctor-skip-self`: Angular NgModule's skipself constructor
- `a-directive`: Directive
- `a-guard-can-activate`: `CanActivate` guard
- `a-guard-can-activate-child`: `CanActivateChild` guard
- `a-guard-can-deactivate`: `CanDeactivate` guard
- `a-guard-can-load`: `CanLoad` guard
- `a-httpclient-get`: `httpClient.get` with Rx observable
- `a-http-interceptor`: Empty Angular `HttpInterceptor` for `HttpClient`
- `a-http-interceptor-headers`: Angular `HttpInterceptor` that sets headers for `HttpClient`
- `a-http-interceptor-logging`: Angular `HttpInterceptor` that logs traffic for `HttpClient`

There are a lot of other client-side framework snippet extensions available as well that can be used to develop applications faster using various technologies and on various platforms.

Visual adjustment extensions

VS Code extensions are divided into various categories. Themes is one of the categories that is commonly used by many developers to adjust the appearance and color of VS Code. This encourages all kinds of developers to choose the theme that suits them.

There are many themes that you can search for from the Extensions bar in VS Code; however, we will be discussing the following three themes in this section:

- Shades of Purple
- Linux Themes for VS Code
- C/C++ Themes

Shades of Purple

Shades of Purple is one of the most popular theme extensions that has more than 725K downloads. You can download it by searching for it in the **Extensions** bar and clicking **Install**.

You can set the theme by hitting the **Set Color Theme** option, as shown in the following screenshot:

Figure 2.10 – Installed and enabled Shades of Purple extension

Linux Themes for VS Code

Linux Themes for VS Code has a set of themes that provide a rich experience of GTK (gnome-tweak-tool) themes. It has more than 189K downloads. You can download it from the **Extensions** bar.

After installing it, you can use the **Set Color Theme** option to set the theme in your VS Code editor.

C/C++ Themes

VS Code supports a variety of platforms and languages. No matter what language or technology we use, VS Code can easily be used to write applications with a similar or even improved development experience compared to any other IDE or editor.

The C/C++ Themes extension is developed by Microsoft and provides a similar experience to what a developer gets when working on C++ editors. You can install it from the **Extensions** bar like the others.

Once it is installed, you can create a new C++ file and get a similar experience to that of a C++ editor:

```
C⁺ HelloWorld.cpp ✕

C⁺ HelloWorld.cpp
1      #include <iostream>
2      using namespace std;
3
4      int main()
5      {
6          cout << "Hello VS Code!";
7          return 0;
8      }
```

Figure 2.11 – Writing C++ code after installing and enabling the C/C++ extension

With visual adjustment extensions, any developer of any platform can work with VS Code to develop applications and make the visual adjustments to match the overall look of VS Code as required.

Productivity extensions

In software development, developer productivity is an important concern. It is vital for the team to make sure that the developers are productive and complete tasks on time. There are various productivity extensions in VS Code that actually help developers make the most of their time to complete their tasks and reach their milestones.

In this section, we will talk about the following three productivity extensions that are useful in software development:

- Visual Studio IntelliCode
- Live Share
- Prettier

Visual Studio IntelliCode

The **Visual Studio IntelliCode** extension provides AI-assisted recommendations as you write code in the Java, Python, JavaScript/TypeScript, and MS SQL languages.

Once this is installed, you can get IntelliSense suggestions as you type, and it helps you to write code faster.

Live Share

Peer programming is one the most challenging practices in software development. In peer programming, two programmers work together on one piece of code. There are also certain scenarios where we need our peers to review or debug/troubleshoot our application in the case of an error.

The **Live Share** extension developed by Microsoft allows developers to share their VS Code projects with other developers in real time. Both parties should have this extension installed, enabling them to join the session and see the code, start debugging sessions, and so on. To install it on your workspace or project, you can open the **Extensions** tab, search for **Live Share**, and hit **Install**.

Once the **Live Share** extension is installed and enabled, restart VS Code to complete the changes. You can then click on the **Live Share** icon from the bottom bar, as shown in the following screenshot, and click on the **Share Now** button:

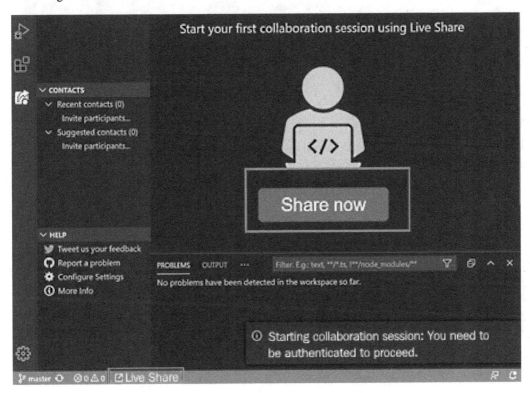

Figure 2.12 – The Live Share icon, after it has been installed

It will ask you to log in with your email account and then, finally, it will show you the link that you can share with your teammate:

Figure 2.13 – Use the Invite page from Live Share to share the session

Prettier

The Prettier extension provides an easy way of formatting code. It supports many languages, such as JavaScript, TypeScript, JSON, CSS, HTML, and others. You can simply add this extension from the **Extensions** pane by searching for `Prettier` and installing it on your VS Code:

Figure 2.14 – Installing the Prettier Extension in VS Code

Once this extension is installed, you can open up the **Settings** tab from VS Code and search for Prettier to list all the configurable settings for Prettier:

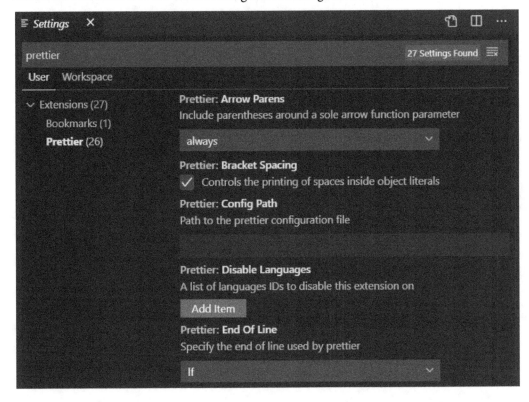

Figure 2.15 – Explore the settings of the Prettier extension

To ensure that the formatting is applied every time you save a file, you can search for the Format On Save setting and click the checkbox, as shown in the following screenshot:

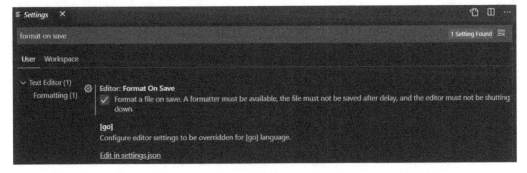

Figure 2.16 – Enabling the Format On Save checkbox to autoformat on saving

Now, once you write any code in a language that Prettier supports, it will perform reformatting when it saves.

Summary

In this chapter, we discussed extensions in detail. We started from the basics of how to search for and install extensions and how to enable and disable them. We also looked at the various extension options that are available to manage and configure them. Extension filters are one of the key features that we can use to filter out extensions based on categories, tags, ratings, and installs. Finally, we covered topics specific to extensions and looked at the various extensions that are useful, such as using code snippet extensions for client-side frameworks, modifying the appearance of VS Code by using various visual adjustment extensions, and enhancing productivity by using Visual Studio IntelliCode, Live Share, and Prettier extensions.

So we have built up our skillset of using VS Code effectively by covering one of the core feature of VS Code—extensions. We learned about the various ways of using extensions that help to increase developer productivity and we are now ready to build applications with VS Code. Therefore, in the next chapter, we will be focusing more on the actual application that we will be building throughout this book and build a backend system based on the Node.js, Java, and Python programming languages.

Section 2: Developing Microservices-Based Applications in Visual Studio Code

This section covers the core part of Visual Studio Code, where we will be learning how to build APIs on different platforms, develop a .NET Core-hosted service to provide integration between different APIs, use Visual Studio Code to develop a client-side web app, develop an app on Azure, and use GitHub with Azure.

This section comprises the following chapters:

- *Chapter 3, Building a Multi-Platform Backend Using Visual Studio Code*
- *Chapter 4, Building a Service in .NET Core and Exploring Dapr*
- *Chapter 5, Building a Web-Based Frontend Application with Angular*
- *Chapter 6, Debugging Techniques*
- *Chapter 7, Deploying Applications on Azure*
- *Chapter 8, Git and Azure DevOps*

3
Building a Multi-Platform Backend Using Visual Studio Code

In the previous chapters, we spent time understanding and getting acquainted with the tool and its important features. To use **Visual Studio Code (VS Code)** for developing enterprise-grade applications, the best way would be to take a use case and go through the complete software development life cycle. For this purpose, we have selected **Job Ordering System (JOS)** as our example.

First, let's talk about the use case we have selected. JOS is an application that will provide users with the ability to request services online. These requests can be related to cleaning, fixing, or anything else. Users can go online, select a particular job type, enter the requested date and time, and submit their request. This request will be automatically picked up by the system and assigned to an agent. The system will also trigger an email notification to the agent.

The frontend of the application will be developed using the Angular framework, while the backend will be developed on a microservices architecture. Job requests by the user will be handled by a Node.js service; agent activities will be managed using a Java service based on the Java Spring Boot framework; the notification service will be developed using Python, and the integration service will be developed using .NET Core. The application will be cloud native, leveraging Azure services.

In this chapter, we will start building the JOS application and first discuss the application architecture. Moving on, we will discuss some key concepts of Azure resource groups and discuss various services part of the cloud-native application architecture. Finally, we will create microservices on different languages and see how VS Code fits well for developing applications as well as for provisioning Azure services.

In line with the preceding introduction, we will cover the following topics in this chapter:

- Overview of our application architecture

- Provisioning managed resources on Azure

- Building a job **application programming interface** (**API**) using Express JS in Node.js

- Building a schedule API in Java using the Java Spring Boot framework

- Building a notification API in Python

Technical requirements

To develop the services in Node.js, Java, and Python, we need to install their respective **software development kits** (**SDKs**). The following are the necessary prerequisites to be installed on the development machine:

- Job API: Node.js

- Schedule API: **Java Development Kit** (**JDK**) version 1.8 or higher; Java Maven version 3.0 or later; and Java Extension Pack

- Notifications API: Python 3.8

Overview of our application architecture

Before we start developing the backend services, let's look at the overall application architecture. The following screenshot gives you a complete overview of what we will be developing and covering in the next few chapters:

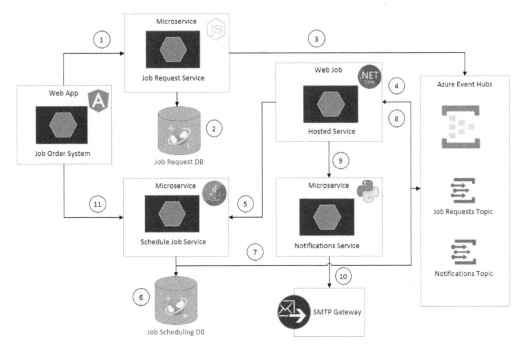

Figure 3.1 – Our application architecture

We can split our application into three main sections, as follows:

- Backend services
- Frontend application
- Integration service

Referring to the application architecture in *Figure 3.1*, the backend microservices are split into the job request service, the agent service, and the notification service. The job request service provides **Create, Read, Update and Delete (CRUD)** operations for the job request database and the agent service provides CRUD operations for the agent database.

The job request service APIs are built using Express JS and they perform database operations on the Azure Cosmos Mongo DB database.

The agent service APIs are built using the Java Spring Boot framework and they perform database operations on their own collection in the Azure Cosmos Mongo DB database.

The notification service is separated to provide an email notification functionality to other services. We will be using the SendGrid API to send email notifications. Encapsulating the SendGrid API in the notification service gives us the flexibility to replace the SendGrid API with any other API in the future, without any impact on the overall application.

To gradually build up our applications, let's keep the discussion limited to the backend services for now. As we move forward, we will start explaining the other components of our application.

We will cover the application development and the services required to be provisioned on the Azure platform. Accordingly, we will be covering the following topics in the upcoming sections of this chapter:

- Building a job API in Node.js using VS Code
- Building a schedule API in Java Spring Boot using VS Code
- Building a notification API in Python using VS Code

While developing these services, we will create the following on Azure:

- A resource group
- An Azure Cosmos DB collection for the job request API
- An Azure Cosmos DB collection for the agent API

Brief introduction to Azure platform services

With a brief introduction of the architecture and our objective for this chapter under our belts, we will start by setting up the Azure platform services required for the backend services.

The focus of this book is on application development with Azure, so we will not elaborate a great deal about the Azure platform, but will go through some important terminologies to make it easier to understand the activities we are carrying out.

Resource group and resources

Resource groups are used to logically distribute resources in Azure. They provide us with the flexibility to manage and segregate our resources. As a good practice, resources linked to a single deployment cycle should be part of one resource group. In the case of shared databases, resource groups can be kept on their own and used by other application components.

Resource groups help in managing and controlling access to resources. Based on your use case and organization, you can create and assign resources to different resource groups. One resource can belong to a single resource group at a time, but it's possible to move them from one resource group to another.

Looking at our example and the size of our application, we have decided to create one resource group and place the backend services, frontend services, and database all in one resource group. You can decide otherwise based on the size of your application. Have a look at the following screenshot:

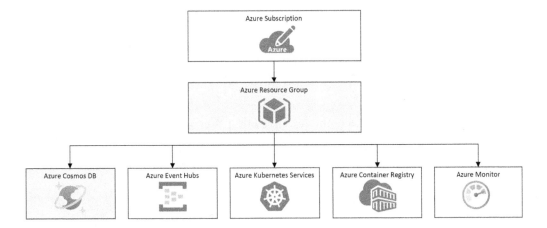

Figure 3.2 – Azure resource group and resources to be provisioned

In *Figure 3.2*, we have put down the complete list of resources we will be provisioning while we continue developing our application throughout the book.

Since this chapter is focused on the backend APIs, we will create our resource group and then provision the Azure Cosmos DB's API for Mongo DB.

Creating an Azure resource group

Azure Resource Manager (**ARM**) is the central service for managing your resources on Azure. The resource manager provides multiple interfaces to create, update, and delete your resources on Azure.

Log in to `portal.azure.com` and click on the **Resource Groups** icon. Alternatively, you can search for **Resource Groups** in the central search bar on the top of the page.

Once you are on the **Resource Groups** page, press the **Add** button and enter the following details:

- **Subscription**: Select the subscription under which you want to create the resource group.

- **Resource Group**: Enter a name of your choice; we have selected VSCodeBookRG as our resource group name.

- **Region**: Select a region. This region specifies where the metadata of your resource group is stored. Resources inside a resource group can exist in the same region or any other region.

Press **Review + Create** and complete the resource group creation.

Finally, you will see your resource group created in the **Resource Groups** page, as shown in the following screenshot:

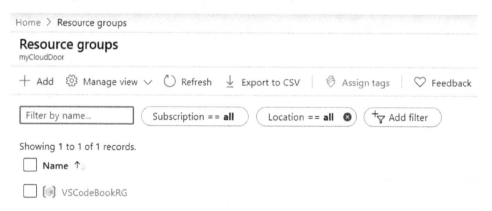

Figure 3.3 – Resource group with the name VSCodeBookRG is created

After creating the resource group, we will start creating the resources inside our newly created resource group.

Since the focus of this chapter is limited to the backend APIs, we will only create the Azure Cosmos DB's API for the MongoDB resource.

Creating Azure Cosmos DB's API for Mongo DB

Similar to the resource group, you can create the Mongo DB resource from the Azure portal. We will also go through the steps of creating the resource using the ARM templates.

Using the Azure portal

To start creating your Mongo DB resource, select the newly created VSCodeBookRG resource group.

Once you are inside the **Resource Group**, press **Add** and search for Azure Cosmos DB, as shown in the following screenshot:

Figure 3.4 – Select the Azure Cosmos DB resource

On the **Create Azure Cosmos DB Account** page, enter the following details:

- **Subscription**: Select the subscription under which you want to create the resource.

- **Resource Group**: Select the VSCodeBookRG resource group created in the previous section.

- **Account Name**: Select a name for the resource; we selected jobinfo.

- **API**: Since we will be using the MongoDB API of Cosmos DB, **select Azure Cosmos DB for Mongo DB API**.

- **Location**: This is the location where your resource will be created, which can be different from your resource group's location.

Press **Review + Create** and complete the resource creation, as shown in the following screenshot:

Home > Resource groups > VSCodeBookRG > New > Azure Cosmos DB > Create Azure Cosmos DB Account

Create Azure Cosmos DB Account

For a limited time, create a new Azure Cosmos DB account with multi-region writes in any region, and receive up to 33% off for the life of your

Select the subscription to manage deployed resources and costs. Use resource groups like folders to organize and manage all your resources.

Subscription *

    Visual Studio Enterprise – MPN   ∨

└──── Resource Group *

    VSCodeBookRG   ∨
    Create new

Instance Details

Account Name *

    jobinfo   ✓

API * ⓘ

    Azure Cosmos DB for MongoDB API   ∨

Notebooks (Preview) ⓘ

    (On   **Off**)

With Azure Cosmos DB free tier, you will get 400 RU/s and 5 GB of storage for free in an account. You can enable free tier on up to one account per subscription. Estimated $24/month discount per account.

Apply Free Tier Discount

    (**Apply**   Do Not Apply)

Location *

    (US) West US   ∨

Account Type ⓘ

    (Production   **Non-Production**)

Version

    3.6   ∨

Geo-Redundancy ⓘ

    (Enable   **Disable**)

Multi-region Writes ⓘ

    (Enable   **Disable**)

*Up to 33% off multi-region writes is available to qualifying new accounts only. Offer limited to accounts with both account locations and geo-redundancy, and applies only to multi-region writes in those same regions. Both Geo-Redundancy and Multi-region Writes must be enabled in account settings. Actual discount will vary based on number of qualifying regions selected.

[**Review + create**]   Previous   [Next: Networking]

Figure 3.5 – The creation form for Azure Cosmos DB resource creation

Azure will take some time to create and deploy your resource; once this is complete, you will see your newly created `jobinfo` resource in the `VSCodeBookRG` resource group.

> **Tip**
>
> For the list of Locations in Azure, you can visit the following URL `https://azure.microsoft.com/en-us/global-infrastructure/geographies/`

Creating the database collections

After creating the Azure Cosmos DB `jobinfo` resource, we will create the collections inside this account.

Go to the `jobinfo` database and create a new collection from **DataExplorer**, or press the **Add Collection** button on the toolbar.

Enter the following details to create the `jobrequests` database:

- **Database Id**: Enter the database ID `jobrequests`.
- **Collection Id**: Enter the collection ID `jobrequests`.
- **Storage Capacity**: Select **Fixed 10 GB**.

Press **OK**, and the database and a collection inside it will be created.

> **Note**
>
> You can also create multiple collections inside the same `jobrequests` database.

Follow the same steps to create the second database and a collection for the agent jobs.

Enter the following details to create the `schedulejobs` database:

- **Database Id**: Enter the database ID `schedulejobs`.
- **Collection Id**: Enter the collection id `schedulejobs`.
- **Storage Capacity**: Select **Fixed 10 GB**.

Using ARM templates

Another way to create the Azure Cosmos DB for Mongo DB API is by using the ARM SDK. For this, we will first need to install the following extensions in VS Code:

- **Azure Resource Manager Tools**
- **Azure CLI** or **Azure PowerShell**

ARM templates help in creating deployment files that can speed up the creation of resources.

Once you have installed the preceding extensions, create an `AzureDeployment` folder and create a new `joborderazuredeployment.json` file inside this folder.

Type `arm`, and VS Code will start showing the snippets for the ARM template, as illustrated in the following screenshot:

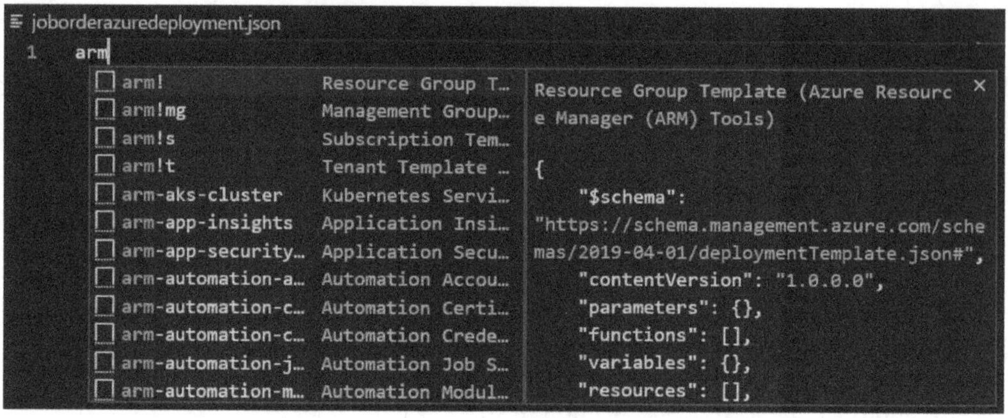

Figure 3.6 – ARM template snippet in VS Code

ARM template for the Azure Cosmos DB resource

Go inside the `joborderazuredeployment.json` file you just created and type `arm`. Press *Enter*, and VS Code will create a basic **JavaScript Object Notation (JSON)** structure for creating a resource, as illustrated in the following code snippet:

```
{
    "$schema": "https://schema.management.azure.com/
    schemas/2019-04-01/deploymentTemplate.json#",
    "contentVersion": "1.0.0.0",
    "parameters": {},
    "functions": [],
    "variables": {},
    "resources": [],
    "outputs": {}
}
```

Installing the Azure CLI

We have already installed the Azure CLI extension earlier. Call the command palette and install the Azure **command-line interface (CLI)**, as shown in the following screenshot:

Figure 3.7 – Install Azure CLI

VS Code will redirect you to the Microsoft download site where you can download the CLI installer for your OS. Download and install the Azure CLI, and then restart VS Code.

In the Terminal window, write `az -version`. If the installation completed successfully in the last step, you will get the CLI version output on the terminal console.

Running the ARM template

To run the commands on your Azure subscription, you first need to log in. Run `az login` to log in to your Azure account.

In case you have not already created the resource group in the preceding section, you can use the following command to create one:

```
az group create --name VSCodeBookRG --location westeurope
```

You can refer to the ARM template script from the Git repository. Here, we will through some sections of the code for understanding. The script will create the Azure Cosmos DB account, Mongo DB database, and the collection.

Let's first define the parameters that will take `Azure Cosmos DB Database ID` and `Collection Id` as the input values, as follows:

```
"parameters": {
    "azcosmosdbacc": {
        "type": "string",
        "metadata": {
            "description": "Azure Cosmos DB"
        }
    },
    "azdatabase": {
        "type": "string",
        "metadata": {
```

```
                    "description": "Database ID"
                }
        },
        "azdbcollection": {
            "type": "string",
            "metadata": {
                "description": "Collection ID"
            }
        }
    },
```

In the `variables` section, we define the locations where the resources will be provisioned, as follows:

```
    "variables": {
        "locations":
    [
        {
            "locationName": "westus",
            "failoverPriority": 0,
            "isZoneRedundant": false
        },
        {
            "locationName": "westeurope",
            "failoverPriority": 1,
            "isZoneRedundant": false
        }
    ]
    },
```

Finally, we specify the resources to be provisioned in Azure. Here, we pass in the parameters entered by the user, the `locations` variables specified previously.

The first resource type is the Azure Cosmos DB database account, as illustrated in the following code snippet:

```
    "resources": [
        {
            "type": "Microsoft.DocumentDB/databaseAccounts",
```

```
        "name": "[parameters('azcosmosdbacc')]",
        "apiVersion": "2019-08-01",
        "location": "westus",
        "kind": "MongoDB",
        "properties":{
            "locations": "[variables('locations')]",
            "databaseAccountOfferType": "Standard"
        }
    },
```

Secondly, we specify the MongoDB database created inside the Azure Cosmos DB database account, as follows:

```
    {
        "type": "Microsoft.DocumentDB/databaseAccounts/
        mongodbDatabases",
        "name": "[concat(parameters('azcosmosdbacc')
        , '/', parameters('azdatabase'))]",
        "apiVersion": "2019-08-01",
        "dependsOn": [
            "[resourceId('Microsoft.DocumentDB/
        databaseAccounts/', parameters('azcosmosdbacc'))]"
        ],
        "properties": {
            "resource": {
                "id": "[parameters('azdatabase')]"
            },
            "options": {
                "throughput": "400"
            }
        }
    },
```

Lastly, we specify the collection to be created in the MongoDB database provisioned in the Azure Cosmos DB database account, as follows:

```
    {
        "type": "Microsoft.DocumentDb/databaseAccounts
```

```
/mongodbDatabases/collections",
            "name": "[concat(parameters('azcosmosdbacc')
, '/', parameters('azdatabase'), '/', parameters
('azdbcollection'))]",
            "apiVersion": "2019-08-01",
            "dependsOn": [
              "[resourceId('Microsoft.DocumentDB/
databaseAccounts/mongodbDatabases', parameters
('azcosmosdbacc'), parameters('azdatabase'))]"
            ],
            "properties": {
                "resource": {
                    "id": "[parameters('azdbcollection')]",
                    "shardKey": { "jobtype": "Hash" }
                },
                "options": {}
            }
        }
    ],
```

The ARM template script can be run using the following command:

```
az deployment group create --resource-group VSCodeBookRG
--template-file joborderazuredeployment.json
```

As shown in the following screenshot, when you run the deployment command, the CLI will ask for three input parameters: the Azure Cosmos DB account (azcosmosdbacc), the database (azdatabase), and the collection (azdbcollection):

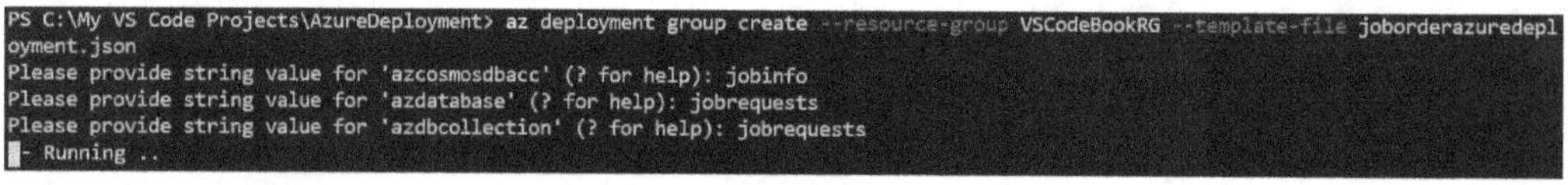

```
PS C:\My VS Code Projects\AzureDeployment> az deployment group create --resource-group VSCodeBookRG --template-file joborderazuredepl
oyment.json
Please provide string value for 'azcosmosdbacc' (? for help): jobinfo
Please provide string value for 'azdatabase' (? for help): jobrequests
Please provide string value for 'azdbcollection' (? for help): jobrequests
- Running ..
```

Figure 3.8 – Running deployment command using Azure CLI

After the script execution is completed, the resources under the VSCodeBookRG group will have been be created.

Follow the same steps to create the agent database and collection, as follows:

- Cosmos DB account name: `jobinfo`
- Database name: `scheduledjobs`
- Collection ID: `scheduledJob`

With the Azure services provisioned, let's move on to developing our first job API in Node.js using VS Code.

Building a job API in Node.js using VS Code

Node.js provides a JavaScript-based environment that runs on the server side. We can develop applications with Node.js using various frameworks. For API development, there are many frameworks available, such as the following:

- Express.JS
- Hapi.JS
- Socket.io
- NestJS
- Feathers.JS

However, we will be developing the job API service using Express.JS. Because of its litheness and speed, Express.JS is one of the most popular frameworks for developing RESTful APIs (where **REST** stands for **representational state transfer**) in Node.js.

Creating a new Express.JS API project in VS Code

VS Code is one of the best tools in the industry for developing applications on Node.js. To start creating a new project in VS Code, we need to first install Node.js on our development machine. You can get Node.js for your OS from `https://nodejs.org/en/download/`. After completing the installation, you can test by running the following command:

```
node --help
```

If Node.js is properly installed, you will see usage and command options on the console. Alternatively, you can also validate if the node is installed by running the following command:

```
node --version
```

Creating a Node.js API project

We will first create a new folder for the job API, and create the Express.JS API project inside it. Open the folder in VS Code and start the terminal window. You can use the **Express Generator Tool** to scaffold the basic Express API template.

Express Generator Tool can be installed as a node module. To install this tool, you can run the following command:

```
npm install -g express-generator
```

The -g switch will make the node module globally available in the machine. Otherwise, you would only be able to use it inside the folder in which you ran the command.

We can then run the following command to scaffold the job API project:

```
express NodeJSAPI
```

Since our API only contains a few CRUD operations, we will just create package. json and server.js files manually on the root folder of our job API project instead of generating them from the tool.

Here is the code snippet of the package.json file:

```
{
    "name": "jobapi",
    "version": "1.0.0",
    "description": "",
    "scripts": {
      "start": "node server.js"
    },
    "author": "",
    "license": "ISC",
    "dependencies": {
      "body-parser": "^1.19.0",
      "express": "^4.17.1"
    }
}
```

In the preceding file, we have provided some metadata information about the API such as name, version, and description. The scripts section specifies the server.js file in the start tag.

To install the dependencies, we need to first run the `npm install` command from the root folder of the application where `package.json` resides, shown as follows:

```
npm install
```

The dependencies section contains the package of Express JS and `body-parser`. Here, the body parser is used to serialize/deserialize objects into JSON format. It is primarily used with HTTP POST requests.

Here is the code snippet of `server.js`:

```javascript
const express = require('express');
const bodyParser = require('body-parser');


// create express app
const app = express();


// parse requests of content-type - application/x-www-form-//
// urlencoded
app.use(bodyParser.urlencoded({ extended: true }));


// parse requests of content-type - application/json
app.use(bodyParser.json());


// define a simple route
app.get('/', (req, res) => {
    res.json({"message": "Welcome to Job API"});
});


// Listen to port 3001
app.listen(3001, () => {
    console.log('listening on 3001')


});
```

In the preceding `server.js` script, we started off by declaring the member variables of the express and body-parser modules. Next, we created an instance of the `Express` app by calling `express()` and initializing the `app` variable. Using the `app` variable, we registered two `middlewares` from the `bodyParser` module. Next, we exposed a simple GET HTTP request on the root path `'/'` that returns a JSON response when `http://localhost:3001` is accessed. Finally, we start the server by calling the `listen` method of the `app` object.

Now, run the application by executing `npm start`. This will open a new browser and hit the root path of the Node.js application. The default port for the Node.js application is `3000`. However, we have changed the port to `3001` using the `app.listen` method, as can be seen in the following screenshot:

Figure 3.9 – JSON output when calling the server root path

In the preceding section, we created a basic Express.JS API project using Node.js that returns the JSON message on an HTTP GET request.

Adding Mongo DB support to the Node.js API

To add MongoDB support, we will add the following dependencies:

```
npm install mongodb -save
```

Once the `mongodb` module is installed, we will modify the `server.js` file and initialize `MongoClient`, shown as follows:

```
//create MongoClient
const MongoClient = require('mongodb').MongoClient
```

Then, add the following code snippet to connect with the Mongo API for Azure Cosmos DB we created in the previous section:

```
MongoClient.connect('mongodb://your_connection_string
', (err, client) => {
    if (err) {
return console.log(err)
    }
    console.log("connected to Cosmos DB Mongo API");
    db = client.
    db("jobrequests"); // whatever your database name is
```

```
app.listen(3001, () => {
    console.log('listening on 3001')


    })
})
```

It will connect to the MongoDB and listen at port `3001` for incoming requests.

To get the connection string for your Azure Cosmos DB account, go to the `jobinfo` resource created on the Azure portal and copy the primary connection string, as shown in the following screenshot:

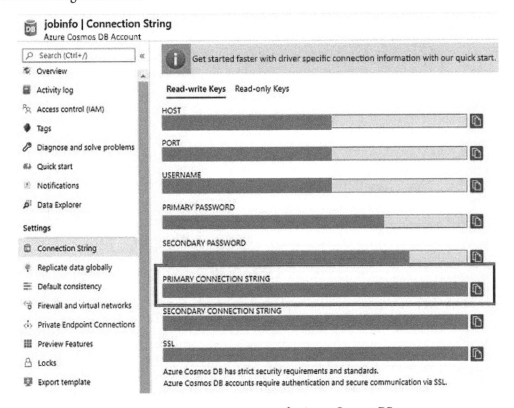

Figure 3.10 – Primary connection string for Azure Cosmos DB account

Adding read and create API methods to the Server JS file

We will now create the `GET` and `POST` methods for the `jobrequests` collection. These methods will be called by our Angular frontend to perform read and save operations on the MongoDB database.

Here is the method for an HTTP GET operation to read the job information:

```
app.get('/jobs', (req,res)=>{
    var cursor = db.collection('jobrequests').find().
toArray(function(err, results) {
        console.log(results);
        res.json("Loaded "+results);
        // send HTML file populated with quotes here
    })
});
```

In the preceding code snippet, using the app object we added a new routing path /jobs to get a job's information. For this, we used the app.get method, passing in the request and response objects as req and res, respectively. Inside the method, we are initializing a database cursor on the jobrequests collection and returning all job requests created so far.

Next, we create an HTTP POST operation to save the job information, as follows:

```
app.post('/jobs', (req, res) => {
    db.collection('jobrequests').insertOne(req.
    body, (err, result) => {
        if (err) {
    return console.log(err);
        }
        console.log('saved to database');
        res.redirect('/');
    })
});
```

Here, we used the post method of the app object. We used the same /jobs routing path and called the insertOne method on our jobrequests collection. Based on the HTTP verb used while making the call from the frontend, either a GET or a POST method will be invoked for the same /jobs path.

For now, our job information API in Node.js is ready to use. We can test it with Fiddler or Postman.

Building a schedule API in Java using VS Code

Java is one of the pioneer programming languages that is based on **Object-Oriented Programming (OOP)** principles. With Java, you can develop enterprise-grade applications, and it provides various frameworks to build an application for your specific needs.

For RESTful APIs, there are a lot of frameworks available, as follows, in which Java Spring Boot is one of the most widely used frameworks:

- Spring Boot
- ACT framework
- Spark
- Dropwizard
- Ninja
- Snow
- Light-rest-4j

In this section, we will build the schedule API in the Java Spring Boot framework with VS Code and show the steps needed to develop, build, and run the application in VS Code.

Creating a new Java Spring Boot API project in VS Code

VS Code offers a very lightweight development environment for developing a Java Spring Boot application. Java Spring Boot provides Maven and Gradle frameworks for developing RESTful applications. With this, we will be using the Maven framework to build our schedule API.

Creating a Java Spring Boot API project

In order to create a new Java API project for our schedule service, we will create a new project directory and name it `JavaSpringBootAPI`. You can give any name for your API project. Open the newly created folder in VS Code.

In order to generate a new Spring Boot API project, we first need to install the Spring Initializer extension. To install this extension, we will first go to the **Extensions** marketplace in VS Code and type `vscode-spring-initializr`. This is a one-time installation on the development machine, and for any future projects we create, we don't need to install this extension again.

After installing the Spring Initializer extension, we can add a new Maven project by opening the **Command Palette** and typing `spring`, as illustrated in the following screenshot:

Figure 3.11 – Generating Maven project using Spring Initializer

This will take you to some wizard steps where you will select `Java` as a project language, `com.jobsystem` as a group ID, `schedule` as an artifact ID, `maven` as the version, and finally, `Spring Web` as the dependencies for your project. Once the project is created, you will see the following files generated:

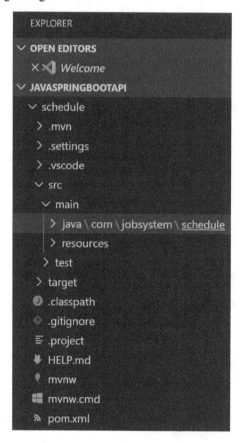

Figure 3.12 – Maven project files generated from Spring Initializer

Adding a schedule controller for the API

Once the Java Spring Boot project is created, we will create a `ScheduleController` class to expose web APIs. To do this, add a new `ScheduleController.java` file in the following folder path: `JavaSpringBootAPI/schedule/src/main/java/com/jobsystem/schedule`:

```java
package com.jobsystem.schedule;
import org.springframework.web.bind.annotation.*;


@RestController
public class ScheduleController {


    @RequestMapping("/")
    public String index() {
        return "Greetings from Spring Boot!";
    }


}
```

In the `ScheduleController` class, we exposed an `Index` method that works on the root path on a default HTTP GET request and returns a greeting message. `@RequestMapping` is the annotation attribute used to specify the routing path, and to use this attribute, we need to add a namespace of `org.springframework.web.bind.annotation` in our class.

> **Hint:**
> You can resolve the namespaces by hovering the cursor on the type and hitting *Command/Ctrl +*.

The default port for the Spring Boot application is `8080`. However, we can change this port by adding the `server.port` value in `application.properties`, as shown in the following screenshot:

```
scheduler > src > main > resources > 🅙 application.properties
  1    server.port = 9003
```

Figure 3.13 – Changing default port for the Java API

So far, we have exposed a simple GET API. To test, hit *F5* in VS Code and start the server. We should be able to see **Greetings from Spring Boot!** once we call `http://localhost:9003` on the browser.

Creating an entity model

Before we create the GET and POST methods for the schedule job API, we will first create an entity class. The data we receive from the HTTP request will be serialized into this entity model.

Here is the `ScheduleJob` class for the entity model discussed in the preceding section:

```
package com.jobsystem.schedule;

import java.util.Date;
import org.springframework.data.annotation.Id;

public class ScheduleJob {

    @Id
    public String id;

    public String jobId;
    public String jobName;
    public String jobType;
    public Date scheduledDate;
    public String agentId;
    public String agentName;


    public ScheduleJob() {}

    public ScheduleJob(String jobId, String jobName,
 String jobType, Date scheduledDate, String agentId,
 String agentName) {
        this.jobId = jobId;
        this.jobName = jobName;
        this.jobType = jobType;
        this.scheduledDate = scheduledDate;
        this.agentId = agentId;
```

```
        this.agentName = agentName;
    }

}
```

The `ScheduleJob` entity class starts by specifying the properties of the class. These properties are the fields in our database collection. The `Id` property is specified as the key field by using the `@Id` annotation. The parameterize constructor of the `ScheduleJob` class takes all the fields as input and assigns them to properties of the `ScheduleJob` object.

Adding the repository and domain model in the schedule API

We will now create a repository to provide CRUD operations for our `scheduledjobs` collection of Cosmos DB using the MongoDB API. Java Spring Boot provides a library that can be used to connect with Mongo DB.

We will modify the `pom.xml` file and add the following dependency in the project:

```
<dependency>
    <groupId>org.springframework.boot</groupId>
    <artifactId>spring-boot-starter-data-mongodb</artifactId>
</dependency>
```

Now, we will create a new `ScheduleRepositorysitory` class and extend this to the `MongoRepositorysitory` interface, specifying the `ScheduleJob` entity model as the type and `String` as the ID.

Here is the code snippet of `ScheduleRepositorysitory.java`:

```
package com.jobsystem.schedule;
import org.springframework.data.mongodb.repositorysitory.
MongoRepositorysitory;


public interface ScheduleRepositorysitory extends
MongoRepositorysitory<ScheduleJob, String>{

}
```

MongoRepositorysitory contains some abstract methods to perform CRUD operations out of the box. However, you can also add more methods to handle other types of operations. For example, if we want to find the schedule by job ID, we can add the following method that takes the jobId as a parameter and returns the ScheduleJob instance as a result:

```
public ScheduleJob findByJobId(String jobId);
```

Adding read and create API methods in ScheduleController

Once the repositorysitory has been set up properly, we can now expose HTTP GET and POST methods in the ScheduleController to create and read job schedule information from the job database.

Here is the HTTP GET method to read the job schedule information:

```
@RequestMapping("/jobs")
public List<ScheduleJob> getJobs(){
    return repositorysitory.findAll();
}
```

And here is the HTTP POST method to schedule the job:

```
@PostMapping("/jobs")
public Boolean saveJob(@RequestBody ScheduleJob job){
    repositorysitory.save(job);
    return true;
}
```

So far, we have created Job and Schedule services. Next, we will develop a Notification service to send out email notifications to the users.

Building a notification API in Python using VS Code

A notification API is used to send an email notification to the person who created the job. The email contains basic job and scheduling information. In order to send the email, we will use the SendGrid resource in Azure. SendGrid is one of the largest cloud-based email delivery platforms that can be provisioned in minutes from the Azure portal.

Creating the SendGrid resource in Azure

To start with, we first create a SendGrid resource in Azure. In order to do this, we have to log in with a valid account that has an active subscription on `portal.azure.com`. Then, we can create a SendGrid resource by creating a new resource in Azure by searching for `SendGrid` in the Azure marketplace, as illustrated in the following screenshot:

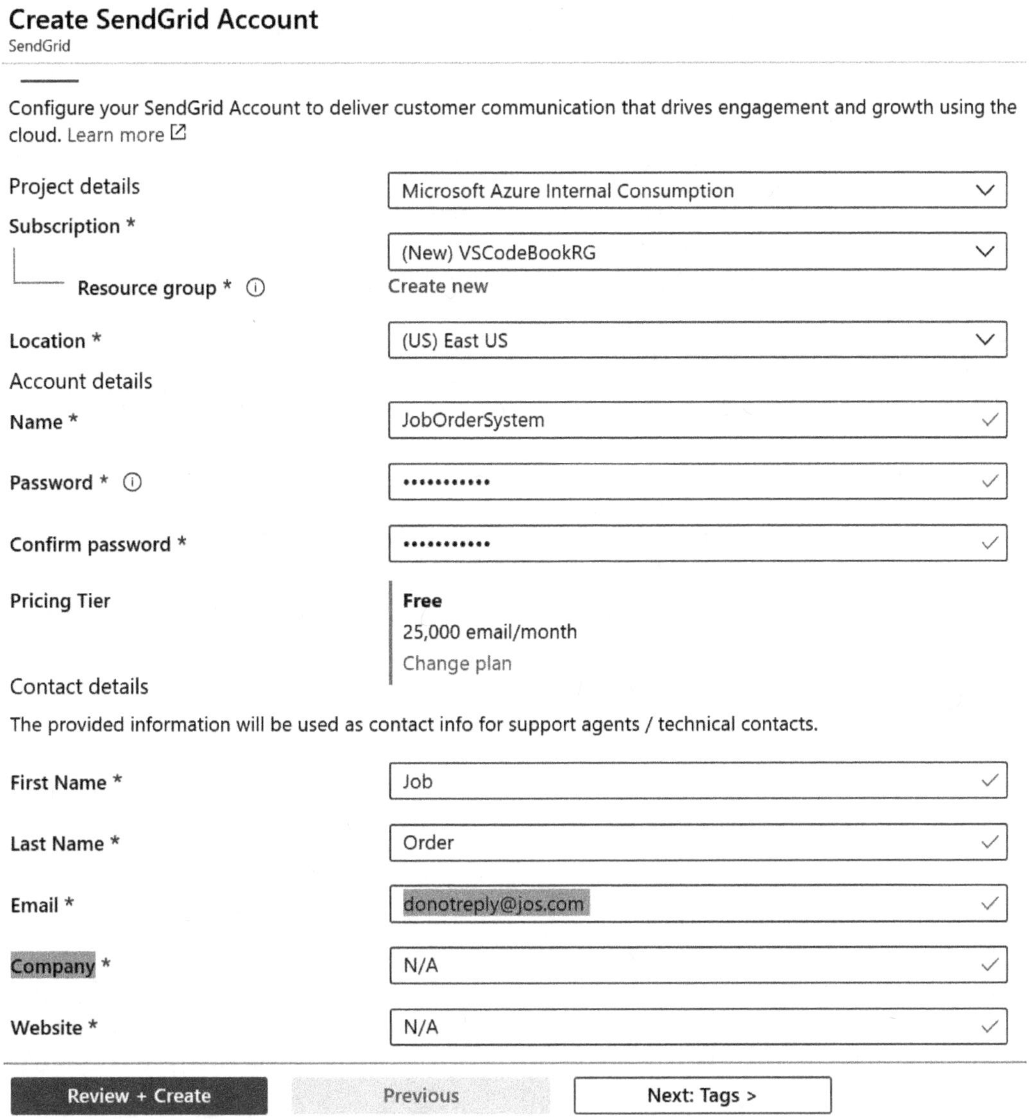

Create SendGrid Account
SendGrid

___

Configure your SendGrid Account to deliver customer communication that drives engagement and growth using the cloud. Learn more ☑

Project details	
	Microsoft Azure Internal Consumption ⌄
Subscription *	
	(New) VSCodeBookRG ⌄
Resource group * ⓘ	Create new
Location *	(US) East US ⌄
Account details	
Name *	JobOrderSystem ✓
Password * ⓘ	•••••••••• ✓
Confirm password *	•••••••••• ✓
Pricing Tier	**Free** 25,000 email/month Change plan

Contact details

The provided information will be used as contact info for support agents / technical contacts.

First Name *	Job ✓
Last Name *	Order ✓
Email *	donotreply@jos.com ✓
Company *	N/A ✓
Website *	N/A ✓

Review + Create	Previous	Next: Tags >

Figure 3.14 – Creating SendGrid resource in Azure

After creating the `SendGrid` resource, click on the **Manage** button, as shown in the following screenshot, to open up the **SendGrid** portal. Here, you can get information on integrating **SendGrid** with the Python API:

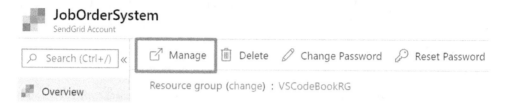

Figure 3.15 – Manage SendGrid resource

From the **SendGrid** portal, we can go inside the **Email API** menu option and click on **Integration Guide**. This will open up a new page on the right pane and show options to integrate based on a web API method or **Simple Mail Transfer Protocol** (**SMTP**) relay. SMTP relay can be used if you want to use the SMTP SDK to integrate with your application. However, we will go with the web API method.

Click on the **Web API** method and choose **Python** as your language to be used, as illustrated in the following screenshot:

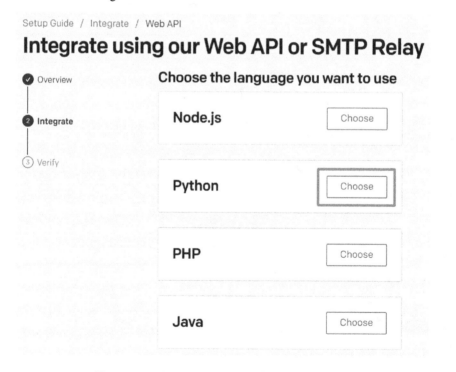

Figure 3.16 – Integration options for SendGrid API

This opens up another page and asks you to specify the API key. You can mention any name for the key and hit **Create Key**. In our case, we used JOS Key, as shown in the following screenshot:

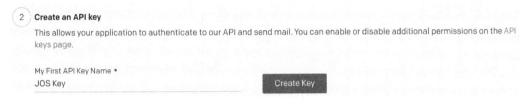

Figure 3.17 – Creating an API Key for SendGrid API

Once the key is created, note it and save it somewhere as we will be using it in our Python API.

Creating a Python API project in VS Code

There are various frameworks for developing web applications in Python. However, Flask is one of the most lightweight frameworks. It provides an easy way to develop applications quickly and define **Uniform Resource Locator** (URL) routing and page rendering easily.

To start with, we will first install the Python extension in VS Code. Open the Python project folder in VS Code and type *Command/Ctrl + Shift + X* to open the extensions bar. Search for the Python extension and install.

From the terminal window in VS Code, run the following command to create a new Python project:

```
python -m venv env
```

This command will create a virtual environment for your current Python interpreter.

Next, we will open the Command Palette using *Command/Ctrl + Shift + P* and select **Python: Select Interpreter**, shown as follows:

Figure 3.18 – Selecting Python Interpreter

Proceed with the wizard and select the version of Python installed on your system. Select the latest interpreter version. After selecting the interpreter, open the Terminal window in VS Code and install the Flask framework. Here is the command to install Flask:

```
pip install flask
```

Once Flask is properly installed, we can add code to develop an API in the Flask framework.

Adding a code file to expose API methods

Now, we will add a new app.py file and expose a method that can be used to send messages using the SendGrid SDK.

We will first check if our application is working fine by adding an HTTP GET method, shown as follows:

```
from flask import Flask, request

app = Flask(__name__)

@app.route("/")
def home():
    return "Hello, Notifications API!"
app.run()
```

In the preceding code snippet, we first import the Flask library from Flask and then initialize it with app variable. We exposed a basic HTTP GET method by adding @app.route method and specified / as the routing path. Finally, we run the application by executing the app.run() method.

To run the application, we can call the Python app.py command, which opens up the port at 5000 and start listening for incoming requests. On navigating to the root URL, we should be able to see the message **Hello, Notifications API** in the browser.

Since the basic API has been set up, let's get started integrating the SendGrid SDK in our Python project. To start, we need to first install the SendGrid SDK. You can do that by executing the following command:

```
pip install SendGrid
```

Next, we will modify the app.py file and import the SendGrid library and expose an HTTP POST method to receive a POST request from the system.

Here is the complete code snippet to use the **SendGrid** library and expose an HTTP POST method for sending notifications:

```python
from flask import Flask, request

import os
from sendgrid import SendGridAPIClient
from sendgrid.helpers.mail import Mail


app = Flask(__name__)

@app.route("/")
def home():
    return "Hello, Flask!"

@app.route('/notifications', methods=['POST'])
def getNotifications():
    message = Mail(
    from_email="donotreply@jos.com",
    to_emails=request.json['toemail'],
    subject=request.json['subject'],
    html_content=request.json['body'])
    try:
        sg = SendGridAPIClient("Enter your SendGrid API key")
        response = sg.send(message)
    except Exception as e:
        print(e)


s
app.run()
```

Finally, you can run this API from the terminal window of VS Code by executing the following command:

```
python app.py
```

You can try to test this API through Fiddler or any other tool. By running the / notifications **Uniform Resource Identifier (URI)** with an HTTP POST request, it will use the SendGrid API and send an email.

Summary

In this chapter, we started off by designing the complete application architecture and moved on to discuss the components we will be developing during this chapter. Once the architecture was in place, we discussed some key concepts of Azure relevant to our project and elaborated on the steps for provisioning our Azure resources. With the Azure resources created, we then next moved on and created the backend services on Node.js, Java, and Python.

We have now learned some basics about Azure resource groups and resources and created backend services in different languages using VS Code.

Next, we will move forward and create a background service in.NET Core while also exploring Dapr.

4

Building a Service in .NET Core and Exploring Dapr

When developing applications based on microservices architecture, the communication between services can be synchronous or asynchronous. Since each service is hosted as a separate process and is exposed over an HTTP endpoint, the integration with other services can be done over an HTTP/HTTPS protocol.

In the case of a distributed transaction, where a single transaction spans to multiple services, calling services synchronously over an HTTP request/response channel is not the best approach. If any service fails, it may fail the whole transaction. Secondly, if each service takes a long time to execute a request, it may result in a request timeout and the transaction would end unexpectedly. There are multiple ways to overcome this challenge, one of which is the message broker technique, which relies on a pub/sub communication model. In the pub/sub model, a source service publishes a message to a queue that is listened to by one or many consumer services that read the message and process it. Alternatively, a middleware service can also be created that subscribes to different queues to listen for messages and provide integration with other services.

In the previous chapters, we created several RESTful services based on the microservices architecture. Each service is tied to a business domain and performs specific tasks. For example, the Node.js service creates a job request, the Java service schedules jobs and assigns agents, and the Python service sends email notifications. As per the flow of the JOS (job order system), once the job is created from the web application, it makes an HTTP request to the Node.js service, which creates the job request in the database and invokes the Java service to schedule a job and assign an agent. Once the agent is assigned, an email notification is sent using the Python service. The whole transaction can be done using a direct HTTP invocation; however, we can decouple this using a message broker technique as an alternative option. In this chapter, we that will be used to listen for events from a queue and provide integration with other services by making a call over HTTP.

The following are the topics that will be covered in this chapter:

- The provision of Azure Event Hubs
- Developing background services in .NET Core to listen for events and provide integration with other services
- Modifying Node.js and Java services to publish messages to Azure Event Hubs over the Kafka protocol
- Exploring Dapr and learning how to integrate with Azure Event Hubs using output bindings

Here, we will begin by understanding how Azure Event Hubs works with Kafka.

Azure Event Hubs with Kafka

Azure Event Hubs is a managed ingestion service that is simple to use for provision and integration with applications. It is very powerful and can quickly handle millions of messages per a second. It also supports the Kafka protocol, which is why client applications that currently rely on Kafka for message broker scenarios can easily shift to Azure Event Hubs without changing their application code.

What is Kafka?

Kafka is an open source software that works on a pub/sub model for storing and reading messages. Kafka persists messages/events published by publishers in a storage method known as topics, and consumers can read/consume those messages by registering to those topics.

Kafka can handle a high velocity and volume of data and serves the user like a high-speed filesystem for commit-log storage. Based on its characteristics, it helps to build large-scale applications that primarily rely on message broker techniques for service-to-service communications or other related scenarios.

> **Note**
> There are various options available in Azure to work with Kafka protocols, some of them including Kafka native cluster, Azure HD Insights, and Azure Event Hubs

Since this chapter is focused on the integration service, we will provision Azure Event Hubs with Kafka to leverage event-driven messaging and establish an asynchronous communication between services. Let's look at how to set up Azure Event Hubs with Kafka in the next section.

Creating Azure Event Hubs

You can create the Azure Event Hubs resource from Azure Portal or through ARM templates.

Using Azure Portal

To start creating your Azure Event Hubs resource, select the JOS resource group *VSCodeBookRG* created in *Chapter 3, Building a Multi-Platform Backend Using Visual Studio Code.*

Once you are inside the resource group, click **Add** and search for **Azure Event Hubs,** as shown in Figure 4.1:

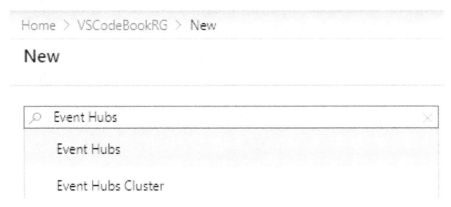

Figure 4.1 – Selecting the Event Hubs resource

On the **Create Azure Event Hubs Account** page, enter the following details:

- **Subscription**: Select the subscription under which you want to create the resource.

- **Resource group**: Select the **VSCodeBookRG** resource group created in the previous section.

- **Namespace name:** Select a name for the resource; we selected `vscodebook`.

- **Location**: This is the location in which your resource will be created—this can be different from your resource group's location.

- **Pricing Tier**: Select the **Standard** pricing plan, since this is required for you to use Event Hubs with the Apache Kafka protocol.

- **Throughputs Units**: Azure Event Hubs traffic is controlled by throughput units. A single throughput unit allows 1 MB per second of ingress and 2 MB per second of egress. With the **Standard** pricing tier, we can configure throughput up to 20 units per second.

Then, click **Review + Create** to complete the resource creation.

Here is a screenshot of the **Create Namespace** page in Azure Event Hubs from Azure Portal:

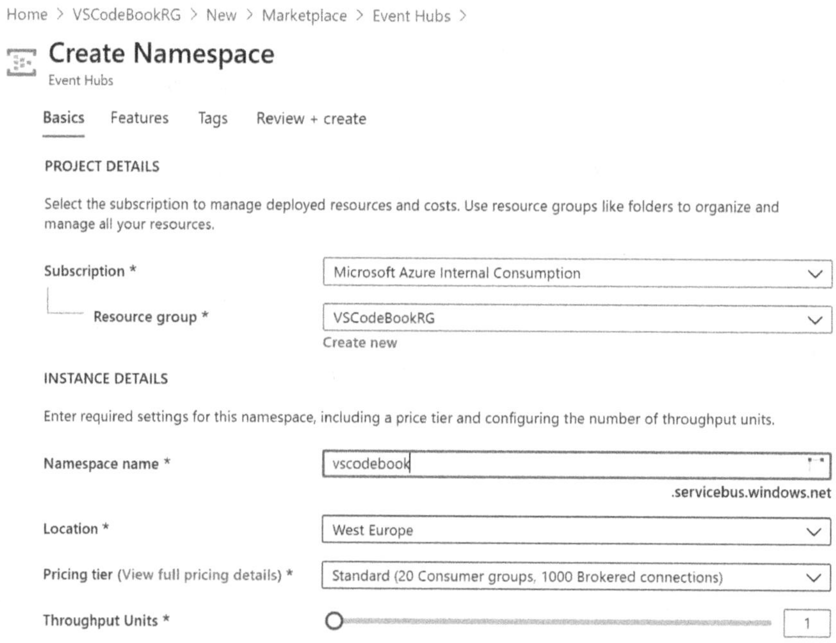

Figure 4.2 – Creating an Azure Event Hubs namespace

Azure will take some time to create and deploy your resource. Once this is complete, you will see your newly created `vscodebook` resource in the **VSCodeBookRG** resource group.

Creating topics in Event Hubs

After creating the Azure Event Hubs namespace, we will create topics to publish messages for both job requests and notification services. From the **Event Hubs** section, you can create topics by selecting the **Event Hubs** option from the **Entities** section.

From **Event Hubs**, click on **Add** to create a new topic for a job request as follows:

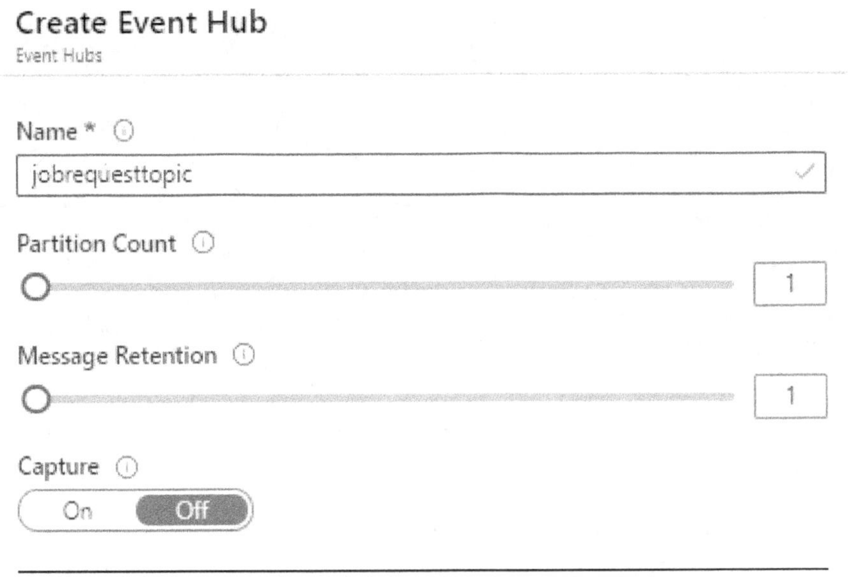

Create Event Hub
Event Hubs

Name * ⓘ

jobrequesttopic ✓

Partition Count ⓘ

○━━ 1

Message Retention ⓘ

○━━ 1

Capture ⓘ

(On   **Off**)

Figure 4.3 – Creating a job request topic in Azure Event Hubs

Here, we have the following fields:

- **Name**: Name of the topic that you can use to refer to this topic.

- **Partition Count**: This denotes the number of concurrent readers you expect to have. Since we will have a single hosted service on .NET Core reading the topics, we can set this to 1.

- **Message Retention**: This is the number of days to keep messages. By default, it is set to 1, but you can increase this if you want to retain messages for a longer period.

Click **Create** to create the topic. Repeat the same steps for the **Notifications** topic, enter the topic name as `notificationstopic`, and keep the rest as their default values.

Using ARM templates

Another way to create Azure Event Hubs is by using the Azure Resource Manager SDK. We have already installed the extension in *Chapter 3, Building a Multi-Platform Backend Using Visual Studio Code*; however, if this is the first time you are working with ARM templates, please refer to *aforementioned chapter*.

Once the extension is installed, create a folder named `AzureDeployment` (or open it if it already exists) and create a new file named `azureeventhubsdeployment.json` inside it.

Type `arm` and VS Code will start showing the snippets for the ARM template:

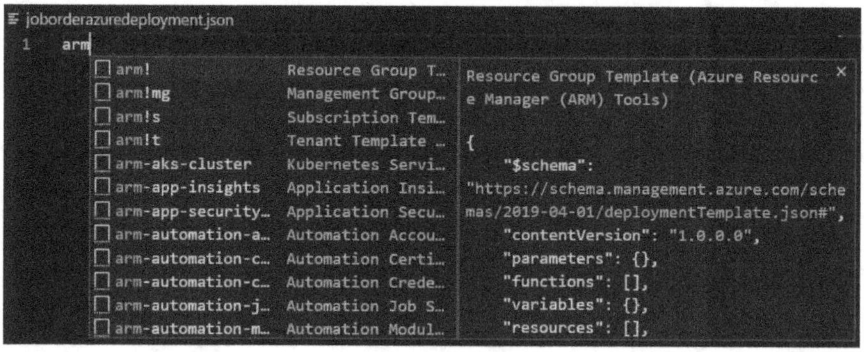

Figure 4.4 – ARM template snippet in VS Code

Figure 4.4 shows the pane with ARM templates. Now, let's create an ARM template for Azure Event Hubs.

ARM template for Azure Event Hubs resources

Go inside the `azureeventhubsdeployment.json` file you just created and type `arm`. Press *Enter* and VS Code will create a basic JSON structure for creating a resource:

```
{
    "$schema": "https://schema.management.azure.com/
    schemas/2019-04-01/deploymentTemplate.json#",
    "contentVersion": "1.0.0.0",
    "parameters": {},
    "functions": [],
    "variables": {},
    "resources": [],
    "outputs": {}
}
```

Here is the script to create the Azure Event Hubs resource with two Event Hub topics, namely, `jobrequesttopic` and `notificationstopic`, respectively:

```json
{
    "$schema": "https://schema.management.azure.com/
    schemas/2019-04-01/deploymentTemplate.json#",
    "contentVersion": "1.0.0.0",
    "parameters": {
        "projectName": {
            "type": "string",
            "defaultValue":"vscodebook1",
            "metadata": {
                "description": "Specifies a project name
                that is used to generate the Event
                Hub name and the Namespace name."
            }
        },
        "location": {
            "type": "string",
            "defaultValue": "[resourceGroup().location]",
            "metadata": {
                "description": "Specifies the Azure
                location for all resources."
            }
        },
        "eventHubSku": {
            "type": "string",
            "allowedValues": [
                "Basic",
                "Standard"
            ],
            "defaultValue": "Standard",
            "metadata": {
                "description": "Specifies the messaging
                tier for service Bus namespace."
            }
        }
    }
```

```json
    },
    "variables": {
        "eventHubNamespaceName":
        "[parameters('projectName')]",
        "eventHubNameForJobRequest":"jobrequesttopic",
        "eventHubNameForNotifications":"notificationstopic"
    },
    "resources": [
        {
            "apiVersion": "2017-04-01",
            "type": "Microsoft.EventHub/namespaces",
            "name": "[variables('eventHubNamespaceName')]",
            "location": "[parameters('location')]",
            "sku": {
                "name": "[parameters('eventHubSku')]",
                "tier": "[parameters('eventHubSku')]",
                "capacity": 1
            },
            "properties": {
                "isAutoInflateEnabled": false,
                "maximumThroughputUnits": 0
            }
        },
        {
            "apiVersion": "2017-04-01",
            "type": "Microsoft.EventHub/namespaces/eventhubs",
            "name": "[concat(variables('eventHubNamespaceName')
            , '/', variables('eventHubNameForJobRequest'))]",
            "location": "[parameters('location')]",
            "dependsOn": [
                "[resourceId('Microsoft.EventHub/namespaces',
                variables('eventHubNamespaceName'))]"
            ],
            "properties": {
                "messageRetentionInDays": 7,
                "partitionCount": 1
            }
```

```
        },
        {
            "apiVersion": "2017-04-01",
            "type": "Microsoft.EventHub/namespaces/eventhubs",
            "name": "[concat(variables('eventHubNamespaceName')
            , '/', variables('eventHubNameForNotifications'))]",
            "location": "[parameters('location')]",
            "dependsOn": [
                "[resourceId('Microsoft.EventHub/namespaces'
                , variables('eventHubNamespaceName'))]"
            ],
            "properties": {
                "messageRetentionInDays": 7,
                "partitionCount": 1
            }
        }
    ]
}
```

Now, run the following command from the PowerShell or VS Code terminal to provision Azure Event Hubs in Azure:

```
az deployment group create --resource-group VSCodeBookRG
--template-file azureeventhubsdeployment.json
```

After the script is executed, the Azure Event Hubs resource will be created under the VSCodeBookRG resource group. It should have both the topics provisioned for job requests and notifications, as follows:

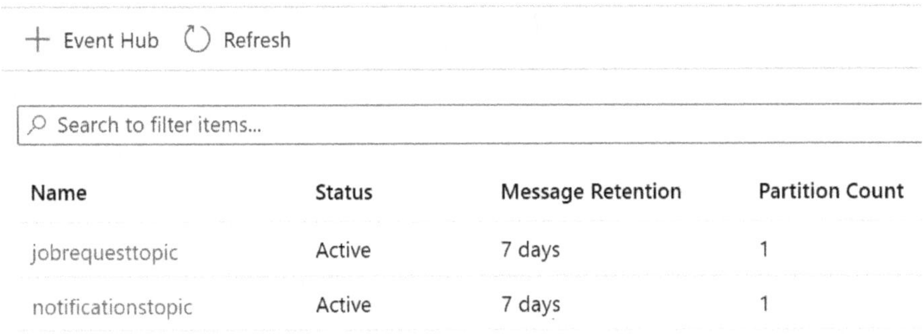

+ Event Hub   ◯ Refresh

🔎 Search to filter items...

Name	Status	Message Retention	Partition Count
jobrequesttopic	Active	7 days	1
notificationstopic	Active	7 days	1

Figure 4.5 – The topics for job requests and notifications in Azure Event Hubs

So far, we have created the Azure Event Hubs resource. In the next section, we will create a background service in .NET Core that listens to Kafka topics of Azure Event Hubs and provides integration with other services.

Building a background service in .NET Core using VS Code

.NET Core is an open source development platform that runs cross platform. It is one of the most widely adopted systems because of its performance. In this section, we will develop a background service in .NET Core that can be used to connect to Azure Event Hubs and listen to Kafka topics. The service makes an HTTP request with job requests and notification services to send messages for the purposes of integration.

The overall scenario is, when the job request is submitted, that the job information is saved in the job request database, and then the job request service publishes an event to the job request topic in Azure Event Hubs. Once the message is published, the .NET Core hosted service picks up the message and calls the schedule job service to schedule the job and assign the agent. Then, once the job has been scheduled, the job scheduler service publishes another event to the notifications topic that can be pulled by the same hosted service and calls the notification service to send out email notifications.

With this approach, the services become loosely coupled and there are no tight dependencies between services. By using Event Hubs and establishing the communication between services using message broker techniques, we can achieve asynchronous communication.

Creating a new .NET Core project in VS Code

.NET Core comes with a **CLI** (short for **command-line interface**) tool that allows you to create projects using commands. To start creating a new project in VS Code, we need to first install .NET Core SDK on our development machine.

> **Tip**
> You can get the .NET Core SDK from `https://dotnet.microsoft.com/download`.

After completing the installation, you can test it by running the following command:

```
dotnet --version
```

Next, we create the project.

Creating a .NET Core API project

We will first create a new folder for the .NET Core API and run the following command to see the .NET Core template available in your system:

```
dotnet new
```

You will see the list of .NET Core templates installed on your system:

```
Templates                                       Short Name            Language       Tags
--------------------------------------------------------------------------------------------------------------
Console Application                             console               [C#], F#, VB   Common/Console
Class library                                   classlib              [C#], F#, VB   Common/Library
WPF Application                                 wpf                   [C#]           Common/WPF
WPF Class library                               wpflib                [C#]           Common/WPF
WPF Custom Control Library                      wpfcustomcontrollib   [C#]           Common/WPF
WPF User Control Library                        wpfusercontrollib     [C#]           Common/WPF
Windows Forms (WinForms) Application            winforms              [C#]           Common/WinForms
Windows Forms (WinForms) Class library          winformslib           [C#]           Common/WinForms
Worker Service                                  worker                [C#]           Common/Worker/Web
Unit Test Project                               mstest                [C#], F#, VB   Test/MSTest
NUnit 3 Test Project                            nunit                 [C#], F#, VB   Test/NUnit
NUnit 3 Test Item                               nunit-test            [C#], F#, VB   Test/NUnit
xUnit Test Project                              xunit                 [C#], F#, VB   Test/xUnit
Razor Component                                 razorcomponent        [C#]           Web/ASP.NET
Razor Page                                      page                  [C#]           Web/ASP.NET
MVC ViewImports                                 viewimports           [C#]           Web/ASP.NET
MVC ViewStart                                   viewstart             [C#]           Web/ASP.NET
Blazor Server App                               blazorserver          [C#]           Web/Blazor
ASP.NET Core Empty                              web                   [C#], F#       Web/Empty
ASP.NET Core Web App (Model-View-Controller)    mvc                   [C#], F#       Web/MVC
ASP.NET Core Web App                            webapp                [C#]           Web/MVC/Razor Pages
ASP.NET Core with Angular                       angular               [C#]           Web/MVC/SPA
ASP.NET Core with React.js                      react                 [C#]           Web/MVC/SPA
ASP.NET Core with React.js and Redux            reactredux            [C#]           Web/MVC/SPA
Razor Class Library                             razorclasslib         [C#]           Web/Razor/Library/Razor Class Library
ASP.NET Core Web API                            webapi                [C#], F#       Web/WebAPI
ASP.NET Core gRPC Service                       grpc                  [C#]           Web/gRPC
dotnet gitignore file                           gitignore                            Config
global.json file                                globaljson                           Config
NuGet Config                                    nugetconfig                          Config
Dotnet local tool manifest file                 tool-manifest                        Config
Web Config                                      webconfig                            Config
Solution File                                   sln                                  Solution
Protocol Buffer File                            proto                                Web/gRPC

Examples:
    dotnet new mvc --auth Individual
    dotnet new console
    dotnet new --help
```

Figure 4.6 – The list of .NET Core templates available to scaffold projects

Then, create a new empty .NET Core web application project by running the following command:

```
dotnet new web --name NetCoreAPI
```

The --name switch is used to specify the project name.

The preceding command scaffolds a new .NET Core web application project. We can use this to open the project in VS Code.

Creating a consumer to listen for Kafka events

To listen to Kafka events, we need to create a consumer that connects to Azure Event Hubs and listens to events. To connect to the Kafka topic, we can use the *Confluent library* that can be added from `NuGet.org`:

- To add the Confluent library, open the command palette in Visual Studio and select **NuGet: Open Gallery**, as shown in the following screenshot:

Figure 4.7 – The selection of NuGet: Open Gallery from the command palette in VS Code

- Once it is opened, search for `Confluent` and install **Conflient.Kafka**, as shown here:

Figure 4.8 – The Confluent Kafka library available in the NuGet gallery

After adding the preceding package, we can now create a consumer to listen for Kafka events.

Creating a consumer class

We will now create a new class that listens for Kafka events by using the Confluent.
Kafka library. Add a new class and name it Consumer.cs:

- To use Kafka with the Confluent.Kafka library, we need to first initialize
  the ConsumerConfig class and specify the configuration values, such as
  the GroupId, BootstrapServers, SaslUsername, SaslPassword,
  SecurityProtocol, SaslMechanism, and Debug properties.

- We will add a new method named GetConsumerConfig. This method sets all the
  values and returns the ConsumerConfig object. Here is the code snippet for the
  GetConsumerConfig method:

```
public ConsumerConfig GetConsumerConfig(){
        var config = new ConsumerConfig
        {
            GroupId = "vscodekafkagroup",
            BootstrapServers = "vscodebook.
            servicebus.windows.net:9093",
            SaslUsername = "$ConnectionString",
            SaslPassword = "",
            SecurityProtocol = SecurityProtocol.
            SaslSsl,
            SaslMechanism = SaslMechanism.Plain,
            Debug = "security,broker,protocol"
        };
        return config;
    }
```

- Next, we will add another method to listen for Job messages. Add a new method and name it ConsumerJobRequestMessages. Here is the code snippet of ConsumerJobRequestMessages:

```
public void ConsumeJobRequestMessages
(CancellationToken cancellationToken)
{
        Console.
WriteLine("Consuming JobRequest messages");
        var config = GetConsumerConfig();
        using (var consumer = new
ConsumerBuilder<Ignore, string>(config).Build())
        {
            consumer.Subscribe("jobrequsttopic");
            var totalCount = 0;
            CancellationTokenSource cts =
 new CancellationTokenSource();
            Console.CancelKeyPress += (_, e) => {
                e.
Cancel = true; // prevent the process from terminating.
                cts.Cancel();
            };
            try
            {                          while
(!cancellationToken.IsCancellationRequested)
                {
                    try
                    {
                        var cr = consumer.
Consume(cts.Token);
                        dynamic data = JObject.
Parse(cr.Message.Value);
                        Console.
WriteLine("Consumed Job Request message");
                        _manager.
CallSchedulerService(cr.Value);
                    }
                    catch (ConsumeException e)
                    {
```

```
                        Console.
    WriteLine($"Error occured: {e.Error.Reason}");
                          }
                      }
                  }
          catch (OperationCanceledException)
              {
                  // Ensure the consumer leaves the
      group cleanly and final offsets are committed.
                  consumer.Close();
              }
          }       }
```

In the preceding code, we start by calling the GetConsumerConfig method that returns the ConsumerConfig object. Then it initializes the ConsumerBuilder object using a using block.

> **Note:**
>
> The using block automatically disposes of objects that were initialized within the block itself.

The Subscribe method of the consumer object is used to subscribe to Kafka events. While using Subscribe, you have to specify the exact topic name. You can validate the topic name from Azure Event Hubs by navigating to the **Event Hubs** tab and matching the name.

To listen to Kafka events, you need to call the Consume method of the Consumer class that actually waits for the events, and once it is pushed to Kafka, it will be triggered and can read the event message. We need to run this method in an indefinite loop; otherwise, once the first message is processed, the program will exit the method. So, we used a while loop to continuously listen to and consume messages from Kafka's jobrequesttopic.

We will implement a similar method for Kafka's notificationstopic. Here is the code snippet for the ConsumeNotificationsMessages method that listens to Kafka notificationstopic in Azure Event Hubs:

```
    public void ConsumeNotificationsMessages
    (CancellationToken cancellationToken)
        {
```

```
        Console.
WriteLine("Consuming Notification messages");
        var config = GetConsumerConfig();
        using (var consumer =
new ConsumerBuilder<Ignore, string>(config).Build())
        {
            consumer.Subscribe("notificationstopic");
            CancellationTokenSource
cts = new CancellationTokenSource();
            Console.CancelKeyPress += (_, e) => {
                e.
Cancel = true; // prevent the process from terminating.
                cts.Cancel();
            };
            try
            {


                while (!cancellationToken.
IsCancellationRequested)
                {
                    try
                    {
                        var cr = consumer.Consume(cts.
Token);
                        Console.
WriteLine("Consumed Notification message");
                        dynamic data = JObject.Parse(cr.
Message.Value);
                        _manager.
CallNotificationService(cr.Value);
                    }
                    catch (ConsumeException e)
                    {
                        Console.
WriteLine($"Error occured: {e.Error.Reason}");
                    }
                }
            }
```

```
        catch (OperationCanceledException)
        {
            // Ensure the consumer leaves
the group cleanly and final offsets are committed.
            consumer.Close();
        }
    }
```

We have created a consumer to listen to messages from Kafka. Now we will develop the background service in .NET Core in the next section.

Creating a hosted service for job requests

Hosted services in .NET Core are used to implement background tasks. A hosted service is a class that inherits the IHostedService interface. We will create two hosted services in a .NET Core API project to listen for both job and notification requests.

To create a new job request hosted service, add a class and name it JobRequestHostedService. The IHostedService interface contains two methods that need to be implemented in the concrete class. Here is the code for IHostedInterface:

```
public interface IHostedService
        Task StartAsync(CancellationToken cancellationToken);
        Task StopAsync(CancellationToken cancellationToken);
    }
```

Any code that is written inside the StartAsync method will be executed when the hosted service is started, whereas the StopAsync method is invoked when the host is performing a graceful shutdown. CancellationToken is used to gracefully shut down the resources and has a timeout duration of 5 seconds. When CancellationToken is requested, all of the background operation applications that are running should be shut down.

Here is the implementation of the job request hosted service:

```csharp
public class JobRequestHostedService : IHostedService
{

    Thread jobRequestServiceThread;
    private Consumer _consumer;
    public JobRequestHostedService(Consumer consumer)
    {
        _consumer = consumer;
    }
    public Task StartAsync
(CancellationToken cancellationToken)
    {
        if(jobRequestServiceThread!
=null && jobRequestServiceThread.IsAlive){
            StopAsync(cancellationToken);
        }
        jobRequestServiceThread
= new Thread(StartJobRequestServiceStart);
        jobRequestServiceThread.Start(cancellationToken);
        return Task.CompletedTask;
    }

    public void StartJobRequestServiceStart
(object cancellationToken)
    {
   try{
            _consumer.
ConsumeJobRequestMessages((CancellationToken)
cancellationToken);
        }
        catch(ThreadAbortException){
        }
    }

    public Task StopAsync
(CancellationToken cancellationToken)
```

```
            {
                if(jobRequestServiceThread!=null){
                    jobRequestServiceThread.Abort();
                }
            return Task.CompletedTask;
            }
        }
```

In the preceding code, we will pass the cancellation token while making a request to
`ConsumeJobRequestMessages`. So, in the case that the service is triggered to be
stopped, the internal loop will end and exit the method.

Creating a hosted service for notifications

Similar to what we did with the `Job Request` service, we will add another class to listen
for notifications. The reason that we create two separate classes for both `Job Request`
and `Notifications` is to run them side by side in separate threads.

So now, we will add a new class named `NotificationHostedService` and add the
following code:

```
public class NotificationHostedService : IHostedService
{
    Thread notificationServiceThread;
    private Consumer _consumer;
    public NotificationHostedService(Consumer consumer)
    {
        _consumer = consumer;

    }
    public Task StartAsync
    (CancellationToken cancellationToken)
    {
    if(notificationServiceThread!=null
    && notificationServiceThread.IsAlive){
            StopAsync(cancellationToken);
        }

        notificationServiceThread
```

```
        = new Thread(NotificationServiceStart);
            notificationServiceThread.Start(cancellationToken);



            return Task.CompletedTask;
        }


        public void NotificationServiceStart
(object cancellationToken){
            try
            {
                _consumer.
ConsumeNotificationsMessages((CancellationToken)
cancellationToken);
            }
            catch(ThreadAbortException)
            {
            }


        }


        public Task StopAsync
(CancellationToken cancellationToken)
        {
            if(notificationServiceThread!=null){
                notificationServiceThread.Abort();
            }
            return Task.CompletedTask;
        }
    }
```

After creating the hosted service, we need to add it to the Startup class so it can be started when the application starts. Let's see how we can add it to the Startup class.

Adding a hosted service to the Startup class

Finally, we can add these hosted services to our `Startup` class in the **ConfigureServices** methods, as shown here:

```
services.AddHostedService<JobRequestHostedService>();
services.AddHostedService<NotificationHostedService>()
```

You can access the code repository to see the complete code.

Modifying the Node.js and Java service to publish messages to Kafka

In the previous section, we created a .NET Core-hosted service to consume messages from Kafka. Now, we need to modify our Node.js API and Java API to publish messages in the following two scenarios:

- When the job request is created from the Node.js API
- When the job is scheduled from the Java API

Let's try it using both of these methods.

Writing Producer in the NodeJS API to publish messages to Kafka

In this section, we will write a code to produce messages to the `jobrequesttopic` of Kafka when the job request is saved in the database:

1. To start setting up `Producer` in the Node.js API, we need to first install the node package, `node-rdkafka`. You can install it by running the following command:

```
Npm install node-rdkafka
```

2. Once the Kafka package is installed, modify `server.js` and add the required package, as follows:

```
var Kafka = require('node-rdkafka');
```

3. Now modify the HTTP `Post` method of `Jobs` and add the following code for `producer` after it:

```
try {
    var producer = new Kafka.Producer({
```

```
    //'debug' : 'all',
    'metadata.broker.list': '',
    'dr_cb': true,  //delivery report callback
    'security.protocol': 'SASL_SSL',
    'sasl.mechanisms': 'PLAIN',
    'sasl.username': '$ConnectionString',
    'sasl.password': ''
});


var topicName = 'jobrequesttopic';

//logging debug messages, if debug is enabled
producer.on('event.log', function (log) {
  console.log(log);
});

//logging all errors
producer.on('event.error', function (err) {
  console.error('Error from producer');
  console.error(err);
});

//
counter to stop this sample after maxMessages are sent
var counter = 0;
var maxMessages = 10;

producer.on('delivery-
report', function (err, report) {
  console.log('delivery-report: ' + JSON.
stringify(report));
  counter++;
});

//Wait for the ready event before producing
producer.on('ready', function (arg) {
  console.log('producer ready.' + JSON.
```

```javascript
stringify(arg));

        var message = new Buffer(jsonPayload);

        // if partition is set to -1,
    librdkafka will use the default partitioner
        var partition = -1;

        producer.produce(topicName, partition, message)

        //need to keep polling for a while to
    ensure the delivery reports are received
        var pollLoop = setInterval(function () {
          producer.poll();
          if (counter === maxMessages) {
            clearInterval(pollLoop);
            producer.disconnect();
          }
        }, 1000);

      });

      producer.on('disconnected', function (arg) {
        console.log('producer disconnected. ' + JSON.
    stringify(arg));
      });
      //starting the producer
      producer.connect();

    }
    catch (e) {
      console.log(e);
    }
```

In the preceding code, we specified `metadata.broker.list` and `sasl.password` as per your `Event Hubs` connection string. The `producer.produce` method is the actual method that sends out the messages to the `jobrequesttopic` in Kafka.

Writing Producer in the Java API to publish messages to Kafka

In this section, we will write a code to produce messages for the `notificationstopic` of Kafka when the job is scheduled, the agent is assigned, and the schedule information is saved in the database. To start setting up the producer in the Java Spring Boot API, we need to first add the dependency for Kafka in the `pom.xml` file:

- Open the Java Spring Boot API project in VS Code and modify the `pom.xml` file by adding the following dependency:

```
<dependency>
    <groupId>org.apache.kafka</groupId>
    <artifactId>kafka-clients</artifactId>
    <version>0.11.0.0</version>
</dependency>
```

- Add the following `SendNotification` method to send messages to Kafka:

```
    private void sendNotificiation
(ScheduleJob job) throws Exception {

        try {
            Properties properties = new Properties();
            // properties.load(new FileReader("src/main/
resources/producer.config"));
            properties.put("bootstrap.servers", "");
            properties.put("security.protocol", "SASL_
SSL");
            properties.put("sasl.mechanism", "PLAIN");
            properties.put("sasl.jaas.config",
                "org.apache.
kafka.common.security.plain.
PlainLoginModule required username=\"$ConnectionString\
" password=\"{Connection String}\";");
            properties.put(ProducerConfig.KEY_SERIALIZER_
```

```java
CLASS_CONFIG, LongSerializer.class.getName());
        properties.put(ProducerConfig.VALUE_
SERIALIZER_CLASS_CONFIG, StringSerializer.class.
getName());


        KafkaProducer<Long, String> producer
= new KafkaProducer<>(properties);
        long time = System.currentTimeMillis();
        Gson gson = new Gson();
        String bidObjectJson = gson.toJson(job);
        final ProducerRecord<Long, String> record =
new ProducerRecord<Long,
String>("notificationstopic", time,
            bidObjectJson);
        producer.send(record, new Callback() {
            public void onCompletion
(RecordMetadata metadata, Exception exception) {
                if (exception != null) {
                    System.out.println(exception);
                    System.exit(1);
                }
            }
        });
        producer.close();
    } catch (Exception ex) {

        }

    }
```

In the preceding code, we specify bootstrap.servers and replace {Connection String} in the sasl.jaas.config property with the actual connection string by referring to our Azure Event Hubs resource. Finally, the producer.send method is used to publish messages to notificationstopic in Azure Event Hubs over the Kafka protocol.

By now, our integration layer is ready. We created a hosted service in .NET Core that listens for Kafka messages and provides integration with other services. In the next section, we will explore Dapr to see how this can be implemented as an alternative option in terms of invoking services or consuming Kafka messages.

Exploring Dapr

Dapr stands for **distributed application runtime**. It's a portable event-driven serverless runtime that helps us build resilient, stateful, and stateless microservices that can run on the cloud or Edge, allowing you to use any language or framework of your choice.

Dapr provides building blocks that can be used with distributed applications. A microservices-based application is one of the right candidates that Dapr can be used with. With microservices, the application is decomposed into a set of fine-grained services that are more modeled toward a business capability or subdomain.

Dapr building blocks

Dapr provides building blocks that can be plugged in with any of your applications to address some of the concerns specific to service invocation, pub/sub communication, state management, and so on. It follows a sidecar architecture and runs as a separate process alongside your app. All the building blocks that Dapr provides are independent of each other, so you can use what you want. We can use any of the building blocks for any technology without adding their dependencies to our application. The following diagram shows the key building blocks of Dapr:

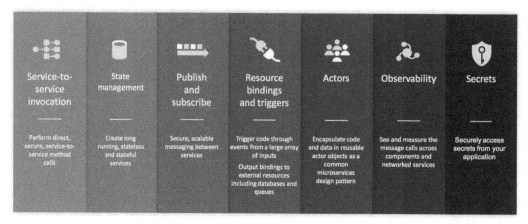

Figure 4.9 – The building blocks of Dapr. The image has been taken from `https://github.com/dapr/dapr`

Service-to-service invocation

Starting from the top-left side, service-to-service invocation is used when we want to make a call from a source to a target service. The target service should run with the Dapr CLI, and the service invocation can be done by following the specific URL pattern:

```
http://<ipAddress>:<daprPort>/v1.0/invoke/<appId>/
method/<methodName>
```

For example, the target service listening on port 3000 with a Dapr sidecar and its jobs method can be invoked by using the following URL pattern:

```
http://localhost:3001/v1.0/invoke/targetservice/method/jobs
```

State management

State management is used to perform state management for storing key–value pairs in any of the storage systems. Any storage system can be plugged in without using its platform-specific libraries in the application code. For example, any storage system, such as Redis Cache, Azure Cosmos DB, and AWS DynamoDB, can be configured in your app following Dapr specifications and can be used over HTTP using the following URL pattern:

```
http://<ipAddress>:<daprPort>/v1.0/state/<stateStoreName>
```

Publish and subscribe messaging

Publish and subscribe messaging is used to implement event-driven messaging based on a pub/sub mechanism, where subscribers subscribe to queues and publishers publish messages on those queues, which are then received by the subscribers. It can be seamlessly integrated with any technology platform, including Azure Event Hubs, RabbitMQ, Amazon Kinesis, and many more.

The following is the URL pattern to publish or subscribe messages to queues:

```
http://{ipAddress}:{daprPort}/v1.0/bindings/{bindingName}
```

Resource

Use resource bindings to trigger an application when events come in or go out to/from external systems. With bindings, you don't need to use any Dapr library within your application code, and with some configuration, it can start listening to events and sending out messages.

Here is the URL pattern to send out messages to any queue as configured in the output bindings configuration file:

```
http://{ipAddress}:{daprPort}/v1.0/bindings/{bindingName}
```

Actors

`Actors` are virtual objects that perform asynchronous computing by encapsulating the state and behavior. They can be used to trigger code execution, make local decisions, send messages, and so on. They run independently with single-threaded execution. When they are idle, they are disposed of through garbage collection. We can use `Actors` to run some background tasks.

Observability

The observability building block can be used to emit logs, traces, and metrics for monitoring purposes. For example, in the case of sending telemetry to the **ApplicationInsights** resource in Azure, we can use this building block to abstract the implementation and use the Dapr SDK to log it to **ApplicationInsights**. In future, if we need to change the underlying logging framework, we don't need to modify the application code.

Secrets

You can plug in any secret stores with Dapr using the secrets building block. For example, you can use Azure Key Vaults to keep an application secret, and instead of using the Azure Key Vault SDK in code, you can use Dapr to store secrets by using the endpoint of `/v1.0/secrets`.

Core components of Dapr

Components are the actual implementation that is used by building blocks. For example, the publisher/subscriber building block uses the publisher/subscriber component, the state building block uses the state component, and so on. The following is the list of components available:

- Service discovery
- State
- Pub/sub
- Bindings

- Middleware
- Secret stores
- Tracing exception

Configuring the Dapr environment

To configure locally, we need to first install the Dapr runtime. The Dapr runtime requires a Docker engine to be installed on the same machine that Dapr needs to be installed on. This means that a prerequisite for Dapr is an installed Docker engine.

Next, we need to install the Dapr CLI to use Dapr and run Dapr commands. To install the Dapr CLI, execute the following command on Windows OS:

```
powershell -Command "iwr -useb https://raw.githubusercontent.
com/dapr/cli/master/install/install.ps1 | iex"
```

The preceding command installs the Dapr CLI and creates a new directory under C:\ Dapr. Also, make sure that the path has been added to the PATH system as well. If not, add an entry of c:\Dapr. This way, you can access the Dapr CLI from any path in your system.

Installing Dapr in standalone mode

Once the Dapr CLI is installed, we can install the runtime using the CLI. The runtime can be installed by running the following command:

```
Dapr init
```

On running the preceding command, it downloads the binaries, sets up the components, and installs Dapr:

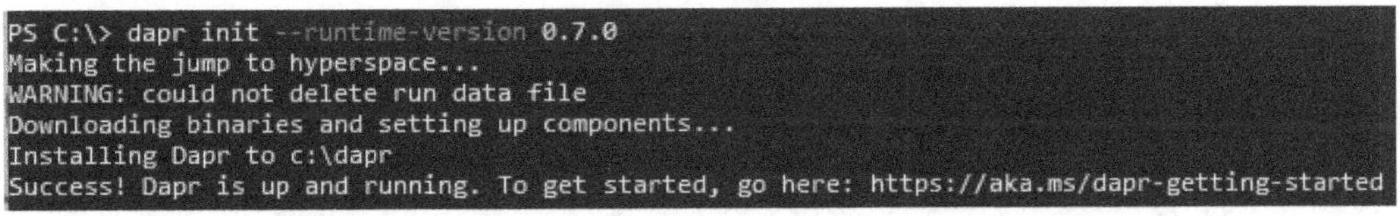

```
PS C:\> dapr init --runtime-version 0.7.0
Making the jump to hyperspace...
WARNING: could not delete run data file
Downloading binaries and setting up components...
Installing Dapr to c:\dapr
Success! Dapr is up and running. To get started, go here: https://aka.ms/dapr-getting-started
```

Figure 4.10 – The initialization of the Dapr runtime installation

> **Note**
> Dapr runs with Linux containers. When installing Dapr, make sure that the Docker engine is switched to Linux containers.

To learn more about the different command options or how to install on a Kubernetes cluster, go to the following link:

```
https://github.com/dapr/docs/blob/master/getting-started/
environment-setup.md#prerequisites
```

Once everything is installed, we can test Dapr by running the `Dapr --version` command.

Service invocation with Dapr

In the previous chapters, we created a few services in Node.js, Java Spring Boot, and Python. Running them alongside Dapr can be done without modifying the application. For example, if we want to run the Node.js API with Dapr, we can run it using the following Dapr command:

```
dapr run --app-id nodeapp --app-port 3001 node server.js
```

In the preceding command, `dapr` triggers the app to run alongside Dapr. `--app-id` is the switch that is used to specify the app name and `--app-port` is the actual port that the app is configured with. We can also specify the Dapr port using the `--port` switch; otherwise, a random port can be assigned after running the command.

Here is the output we see after running the preceding command:

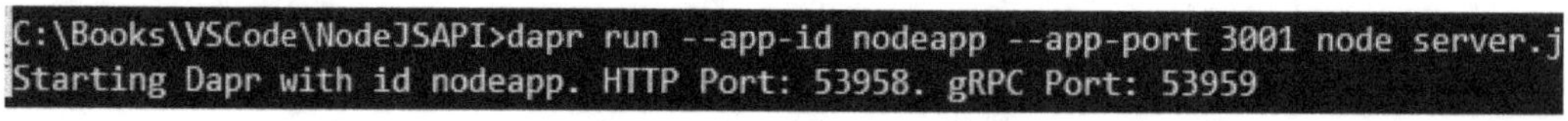

```
C:\Books\VSCode\NodeJSAPI>dapr run --app-id nodeapp --app-port 3001 node server.j
Starting Dapr with id nodeapp. HTTP Port: 53958. gRPC Port: 53959
```

Figure 4.11 – The command to run a Node.js application alongside Dapr

We can see that port **53958** has been assigned dynamically after running the `dapr` command. Dapr provides a special URL pattern to invoke services running with Dapr.

Here is the URL pattern for service invocation:

```
http://<ipAddress>:<daprPort>/v1.0/invoke/<appId>/
method/<methodName>
```

Following the previous URL pattern, the Node.js API `Jobs` method can be invoked as follows:

```
http://localhost:53958/v1.0/invoke/nodeapp/method/jobs
```

Similarly, to run the Java API using Dapr, we can run it using the following command:

```
dapr run --app-id schedulerapp --app-port 9003  -- mvn spring-
boot:run
```

In the preceding example, we also specified the Dapr port so that a dynamic port will not be assigned at runtime:

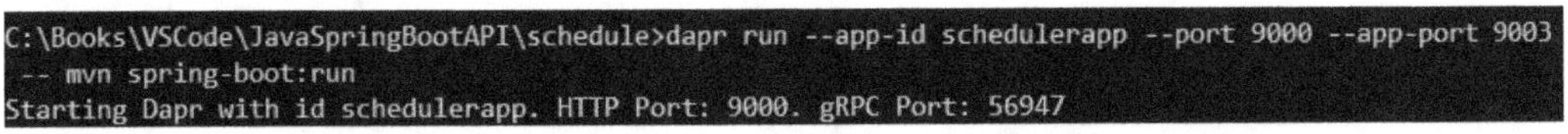

```
C:\Books\VSCode\JavaSpringBootAPI\schedule>dapr run --app-id schedulerapp --port 9000 --app-port 9003
-- mvn spring-boot:run
Starting Dapr with id schedulerapp. HTTP Port: 9000. gRPC Port: 56947
```

Figure 4.12 – Showing the command to run Java API alongside Dapr

Lastly, you can access the Java service using a similar URL pattern, as follows:

```
http://localhost:9000/v1.0/invoke/schedulerapp/method/jobs
```

We have seen how Dapr provides a standard interface to invoke services, as shown in the preceding code. Any consumer or client application can easily consume services by following the standard URL pattern if the services are running alongside Dapr.

Pub/sub communication with Dapr

So far, we have learned how to set up Dapr and run applications with Dapr. In the JOS, we have an event-driven scenario where once the job is created from the web frontend application, it communicates to the Job API to save the job request and then publishes a message to the job request topic. This triggers the hosted service to read the message and then send the message to the schedule service to schedule the job and then publishes another message to the notifications topic. Finally, the notification thread listening to the notifications topic in the hosted service is triggered and calls the Notifications API to send out notifications.

To send messages to Azure Event Hubs using the Kafka protocol, we used the Kafka libraries. With the Dapr pub/sub building block, we can remove the dependencies of the Kafka libraries to publish messages to Azure Event Hubs. This can be achieved either by using the input/output bindings and configuring the files to abstract the pub/sub communication, or by using Dapr SDK. We will explore the output binding and integrate it with both the Node.js and Java API to send out messages to Azure Event Hubs. Let's see how this works in the following sections.

Bindings in Dapr

To leverage the pub/sub component of Dapr, we can either use the SDK or libraries of Dapr or configure input or output bindings to invoke external resources, such as Azure Event Hubs and Service Bus.

In our case, in order to publish messages to the Azure Event Hubs job request topic, we can configure the output binding instead of using Kafka libraries in the Node.js API and invoke the service over HTTP. When the application is run with Dapr, the components folder is created at the root of the application that the command was run from.

To connect with Event Hubs, we need to create the output binding file and place it inside the components folder.

Here is the content of the output binding file named `eventhubs_binding.yaml`:

```yaml
apiVersion: dapr.io/v1alpha1
kind: Component
metadata:
  name: eventhubsoutput
spec:
  type: bindings.kafka
  metadata:
    - name: brokers
      value: vscodebook.servicebus.windows.net:9093
    - name: publishTopic
      value: jobrequesttopic
    - name: consumerGroup
      value: $Default
    - name: authRequired
      value: true
    - name: saslUsername
      value: $ConnectionString
    - name: saslPassword
      value: {ConnectionString}
```

The name property is used to give any name to the binding. As we can have multiple bindings and the name is part of the URL pattern, we will use the binding over HTTP. The type should be bindings.kafka; this is required to connect with Azure Event Hubs over the Kafka protocol. The rest of the metadata, such as brokers, publishTopic, consumerGroup, authRequired, saslUsername, and saslPassword, should be specified as per the actual values for your topic in Azure Event Hubs.

Once this is configured, we can run the NodeJS API using the Dapr CLI, as shown in the preceding code, and make a POST request to the binding's endpoint. To invoke output bindings, we can make a HTTP POST request and use the following URL pattern:

```
http://{ipAddress}:{daprPort}/v1.0/bindings/{bindingName}
```

Here is the actual invocation of the bindings for the Node.js API:

```
http://localhost:3000/v1.0/bindings/eventhubsoutput
```

Note that eventhubsoutput is the actual name given when we configure the bindings in the YAML file. You can test the preceding endpoint by making an HTTP POST request from Fiddler or any other tool and send a message in a request body. If everything is configured correctly, you should receive the HTTP 200 response code and the message will be published to the Event Hubs topic.

We can also modify the Node.js API to make a call to this endpoint once the job request information has been saved in the database. Here is the code snippet to make an HTTP POST request from the Node.js API:

```
// Sending through DAPR
const PORT = process.env.DAPR_HTTP_PORT;
var url = "http://localhost:"+PORT+"/v1.0/bindings/
eventhubsoutput"
        console.log("URL is " + url);
        axios.post(url, { data: req.body })
          .then(res => {
            console.log(`statusCode: ${res.statusCode}`)
            console.log(res)
        })
        .catch(error => {
            console.error(error)
        })
```

With this, we have covered a major part of Dapr and learned about its applications.

Summary

In this chapter, we started off by discussing the ways of establishing communication between services. We discussed the use of the message broker technique to achieve asynchronous communication and used Azure Event Hubs with the Kafka protocol.

We set up Azure Event Hubs using the ARM template and created two topics for job requests and notifications, respectively. We developed a background service in .NET Core using the .NET Core hosted service model and used the Confluent library to connect with Azure Event Hubs and listen to Kafka topics. Then, we modified the Node.js and Java APIs to support publishing events to respective Kafka topics.

Lastly, we explored how Dapr can provide us with various building blocks to accommodate different scenarios and looked at an alternative way of communicating to other services running alongside Dapr and publishing messages using output binding.

In the next chapter, we will develop a frontend SPA (single-app application) in Angular and learn about how to effectively use Visual Studio Code to develop web applications in the Angular framework.

5
Building a Web-Based Frontend Application with Angular

In the previous chapters, we completed our multiplatform backend on Node.js, Java, Python, and .NET Core. These backends have exposed the required endpoints that will be used by the frontend to create and update data. To develop our frontend application, we have selected Angular, and this chapter will focus on taking you through the development of our web application's frontend using **Visual Studio Code (VS Code)**.

Since the focus of our book is on utilizing VS Code, we will not be able to go into a lot of detail about Angular, but we will make sure that if you follow the chapter, you will end up creating the frontend application.

Whatever is used to develop the frontend application will be explained in reasonable detail so that you can follow along and experience developing the application in VS Code. As you will have understood by now, our objective is to make you comfortable with the development environment by taking you through a complete journey, and, as they say, *practice makes perfect*. The more you develop with VS Code, the more comfortable you will get.

To cater to a wider audience, we will start off with an introduction to Angular and gradually unwind the framework features and how they are used, in a real-world example. We will be using Bootstrap to style our app, so this is also something we will cover. This chapter will try to accustom you to the Bootstrap documentation and how you can use it for a particular use case.

The following topics are covered in this chapter:

- A quick overview of Angular
- Designing our application
- Setting up the environment for developing an Angular app using the Angular **command-line interface (CLI)**
- Styling the application using Bootstrap
- Creating the landing page
- Creating the feature modules
- Understanding routing and lazy loading for better app performance
- Authorizing applications
- Developing the components for each feature module

Technical requirements

Before we start developing the application, let's go through the steps for setting up the environment.

Installing Node.js

The first step to set up the environment for developing an Angular app is to install the Node.js framework. You can do this by downloading the required version from `https://nodejs.org/en/download/`.

To check the existing node version, run `node -v` in a terminal window.

The Node.js installation will install the npm (Node Package Manager) CLI. To check the npm version, run `npm -v` in a terminal window.

Installing the Angular CLI

Angular provides a CLI to speed up the development of an Angular application. This CLI helps in quickly generating the app structure, modules, components, services, and pipes, among other aspects.

Since we will be using the CLI to develop this application, let's first install the CLI using the recently installed npm. Open a terminal window and run the following command:

```
npm install -g @angular/cli
```

A quick overview of Angular

Angular is a **Single Page Application** (**SPA**) framework for developing web apps and writing in TypeScript. TypeScript is a typed language that complies into JavaScript—it's a superset of JavaScript.

Angular applications are organized into **modules**. Any Angular app should have at least one module. Modules contain **components** and can also import other modules. This provides a better way to organize your app into reusable blocks of code. A component attaches a **HyperText Markup Language** (**HTML**) template, which provides the methods and attributes to support the view and also handles navigation.

Services help in separating tasks that can be used across an application by several components. This can be an authorization service to authenticate/authorize users, or data services to fetch data from the application backend. Services are injectable classes that can be injected by any component of an application.

Angular follows a simple decorator concept to distinguish between a class of type module, component, or service. A class with an `@NgModule({})` decorator defines a module; `@Component({})` defines a component; and `@Injectable({})` defines a service. Each of these decorators has a set of parameters passed in between the curly braces.

Applications' routing is managed by maintaining routing paths in each module. These route paths are called by the components or services to navigate to a particular component. Since Angular is a SPA, there is always one active page in the browser, and on each navigation, the existing page is replaced by the target component's HTML template.

Angular provides a set of modules that can be imported by our application modules to leverage on the existing libraries. You can also import your own created modules into other modules for reusability and to better organize your applications.

This was just a quick overview of Angular; however, you can refer to the Angular documentation at `https://angular.io` for further details.

Designing our application

In *Chapter 3, Building a Multi-Platform Backend Using Visual Studio Code* we created an overall application architecture. The architecture highlighted the services, integration platform, and the frontend application to be built, and how they work together. From this architecture, we have already developed the backend services for job request, agent, and notification functions. Our focus now will be on developing the frontend application and on providing users with the ability to create job requests, for agents to process them.

To start developing the required Angular modules, components, and services, let's look at an illustration of how we will be organizing our app into *different modules*, showing which components will be part of each module and which prebuilt Angular modules we will be reusing. This will help us to understand the overall application and will also make it easy to link things while we develop. The following screenshot shows how the Angular app is broken down into modules:

Figure 5.1 – Frontend application design

The modules help organize our app in terms of features and serve as the basis for lazy loading. **Lazy loading** helps in increasing the load performance of the app. You will learn more about lazy loading when we work to configure routing in the *Creating the feature modules* section. The modules are detailed as follows:

- **App Module**: `AppModule` is the root module. It bootstraps the `AppComponent` module, which holds the `router-outlet` tag. As we discussed earlier, Angular is a SPA framework. This means that the app does not navigate to a new page; instead, the HTML template of the target component is replaced in the `AppComponent` HTML template. This space is defined in the `AppComponent` HTML by using the `router-outlet` tag.

  To show the HTML template generated by the Angular app to the user, the `AppComponent` selector is placed in the `index.html` file.

- **Login Module**: The app will be divided into three feature modules. The `Login Module` will define the `LoginComponent`, to provide login functionality by entering the user ID and password. The `LoginComponent` will use the `AuthorizationService` to authenticate and authorize the user.

- **Job Request Module**: `JobrequestModule` will define `JobrequestComponent` to create job requests, `JobrequestlistComponent` to show the list of created jobs, and `JobrequestdetailComponent` to show the details of each job. These components will use the `JobrequestService` module to access the backend HTTP services built using Node.js.

- **Agent Module**: `AgentModule` will define `AgentlistComponent` to show the list of assigned jobs for the agent, while `AgentJobprocessComponent` will allow the agent to see assigned job details and the status of the update. These components will use the `AgentService` module to access the backend HTTP services built using *Java Spring Boot*.

Creating and running the Angular app project

With the environment set up for development, we will now move forward and start using the CLI to create and run the app.

Generating the app using the CLI

To generate the Angular app using the Angular CLI, run the following command in a terminal window:

```
ng new JobOrderSystem --routing --prefix jos --style css
```

Here, ng is a keyword, followed by new to create a new project with the name JobOrderSystem. The --routing option will create AppRoutingModule and add the basic routing features. We define a prefix for the component selectors by using the --prefix jos addition, as illustrated in the following screenshot:

```
C:\My VS Code Projects>ng new JobOrderSystem --routing --prefix jos --style css
CREATE JobOrderSystem/angular.json (3630 bytes)
CREATE JobOrderSystem/package.json (1292 bytes)
CREATE JobOrderSystem/README.md (1031 bytes)
CREATE JobOrderSystem/tsconfig.json (489 bytes)
CREATE JobOrderSystem/tslint.json (3125 bytes)
CREATE JobOrderSystem/.editorconfig (274 bytes)
CREATE JobOrderSystem/.gitignore (631 bytes)
CREATE JobOrderSystem/browserslist (429 bytes)
```

Figure 5.2 – Angular app generation using the Angular CLI

The preceding screenshot shows the Angular app being generated by running the Angular CLI command.

Looking at the CLI-generated files

Let's have a look at the CLI-generated files, as follows:

- src: Inside the src folder, the app folder will contain the Angular artifacts.
- app.module.ts: This is the AppModule. Every component in Angular needs to be part of a module. The AppModule will declare the AppComponent and will also bootstrap it.

  It also imports the Angular BrowserModule module, which is required to run our app in the browser. Since we ran the ng new command with the --routing addition, the CLI also generated an AppRoutingModule component. This module will hold the application routes.

- `app.component.ts`: This is the main `AppComponent`. It defines a `jos-root` selector, `templateUrl`, which points to `app.component.html`, and `styleUrl`, which points to the `app.component.css` file.

- `index.html`: This file specifies the `jos-root` selector of the `AppComponent`, as shown in the following code snippet:

```
<body>
    <jos-root></jos-root>
</body>
```

Running the app using the CLI

The CLI generates a complete project scaffolding and is ready to run. To compile and run the **Angular Live Development Server**, use the following command:

```
ng serve -o
```

The -o addition will open the browser once the app is compiled and running. Any changes you make to the app will recompile and update the app in the browser. By default, the app will be running on `http://localhost:4200`.

Adding styling with Bootstrap

Bootstrap is one of the most famous toolkits for creating a responsive website. For more information, visit `https://getbootstrap.com/`.

Installing the Bootstrap and jQuery packages

To install these, we perform the following steps:

1. To add Bootstrap styling to our project, we will need to install the `bootstrap` package. We will do this by using npm.

2. Run the following command in the project directory to install the `bootstrap` package and also add it to the `package.json` file:

```
npm install bootstrap
```

3. Next, add the `jquery` package by running the following command in the same directory:

```
npm install jquery
```

4. You can also run these commands together by running the following command:

```
npm install bootstrap jquery
```

The packages are now downloaded. Next, let's look at the steps to configure the Angular app project.

Adding the CSS and JS files to the Angular project

For the Angular project to access the Bootstrap and jQuery files, we will add them to the `src/angular.json` file, as follows:

```
'styles': [
    'src/styles.css',
    'node_modules/bootstrap/dist/css/bootstrap.min.
    css'
],
'scripts': [
    'node_modules/jquery/dist/jquery.min.js',
    'node_modules/bootstrap/dist/js/bootstrap.min.js'
]
```

In the preceding configuration, we added the Bootstrap **Cascading Style Sheets** (**CSS**) file path in the `styles` array and the JavaScript file paths for both Bootstrap and jQuery in the `scripts` array.

Creating the landing page

The project is set up, and it's time to get on with developing the components and services. Every application should have a landing page that serves as the entry point for your app. If you recall the application design, we will be using the `HomeComponent` module to define the home page.

Generating the HomeComponent

The Angular CLI supports the generation of all types of Angular artifacts. To generate the home component, use the following command:

```
ng generate component home
```

The preceding command will create a home folder and generate the component files in it. Since every component should be part of a module, the CLI will also update the `AppModule`.

> **Note**
>
> While generating the Angular objects, another handy option is to use the −d addition, like this: `ng generate component −d` The CLI will then show the list of objects to be generated or updated without actually creating or updating them.

Setting the application's default route path

We already know that the `HomeComponent` will serve as the application's landing page. To automatically route the user to the home page, we will follow these next steps:

1. Define the route path for the `HomeComponent`.

In the `app-routing.module.ts` file, the home path points to the `HomeComponent`, as illustrated in the following code snippet:

```
{ path:'home', component: HomeComponent },
```

2. Set home as the default path. We use `redirectTo 'home'` when the **Uniform Resource Locator (URL)** application is called with no path, as follows:

```
{ path: '' , redirectTo: 'home', pathMatch:'full'},
```

> **Note**
>
> In a case where the user enters a path that is not defined, we can redirect the user to an error page. For this, use the `path: '**'` wildcard path.
>
> Please note that the sequence of the route path is important. The default path and the wildcard should be placed after the application route paths.

Since `AppComponent` is the root component for our applications, we will need to add the place where our `HomeComponent` template can be displayed. For this, we add a `router-outlet` tag in the `app.component.html` file.

Remove all the code in the `app.component.html` file and place the following code within it:

```
<h1>App Component</h1>
<div class='container'>
  <router-outlet></router-outlet>
</div>
```

The `router-outlet` tag is replaced by the active route. In our case, since the default path is home, `HomeComponent` will be loaded and `home.component.html` will replace the `router-outlet` tag.

To test the app while we develop, split the terminal window and run `ng serve -o`. You will get the following screen:

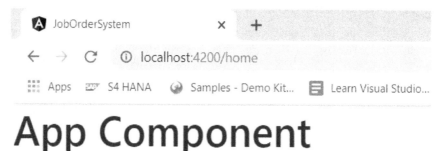

Figure 5.3 – Angular app loaded and redirected to the default route path

The preceding screenshot shows that adding the `-o` addition to `ng serve` launches a new browser window and loads the application. You will also notice that the app will directly route to the `/home` path; this is because of the default route path defined in the preceding paragraph.

> **Note**
> The default port used by Angular is `4200`.

Creating the navigation bar

We have already hooked up the Bootstrap files in our Angular app configuration. We will use the Bootstrap toolkit to style our application and give it a nice look and feel. Let's start by defining the navigation bar for the application, as follows:

1. Go to `https://getbootstrap.com/` and visit the **Documentation** link. In the **Components** side menu, select **Navbar**, then scroll down to the **Nav** section and copy the code, as illustrated in the following screenshot:

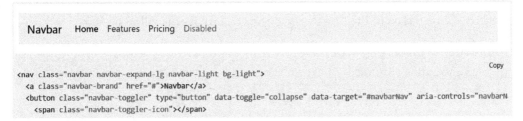

Nav

Navbar navigation links build on our `.nav` options with their own modifier class and require the use of toggler classes for proper responsive styling. **Navigation in navbars will also grow to occupy as much horizontal space as possible** to keep your navbar contents securely aligned.

Active states—with `.active`—to indicate the current page can be applied directly to `.nav-link`s or their immediate parent `.nav-item`s.

Navbar	Home Features Pricing Disabled

```
Copy
<nav class="navbar navbar-expand-lg navbar-light bg-light">
  <a class="navbar-brand" href="#">Navbar</a>
  <button class="navbar-toggler" type="button" data-toggle="collapse" data-target="#navbarNav" aria-controls="navbarN
    <span class="navbar-toggler-icon"></span>
```

Figure 5.4 – Bootstrap site Navbar section

The preceding screenshot shows the Bootstrap documentation page, with an option to copy the code snippet.

2. Paste the copied code into `app.component.html`, and place JOS in place of Navbar.

    Our application is already running, so switch to the browser to see the updated page:

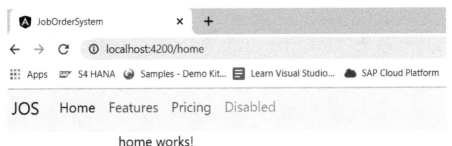

Figure 5.5 – Navbar added to the application

The preceding screenshot shows **Navbar** added to the application.

Designing and styling the home page

Let's work to make the home page more attractive. Going back to the Bootstrap toolkit, this time we will use the **Carousel** control and place a few images.

In the Bootstrap **Components** section on getbootstrap.com/docs/4.5/ components, go to **Carousel** and select the one you like.

Copy the code and paste it into the home.component.html file. Select a few images and place them in the assets folder. Assign the images in the src attribute for all img tags, as follows:

```
<img src='/assets/img1.jpg' class='d-block w-100' alt='...'>
```

To reduce the height of the carousel on your page, set the max-height attribute of the carousel-inner class in the home.component.css file, as follows:

```
.carousel-inner{
    max-height: 500px;
}
```

Use the Bootstrap **Card** component to place some images and text. For this, visit https://getbootstrap.com/docs/4.5/components/card/ and copy the **Card** HTML template. Use the one with an image and some text below. Add a main div tag after the carousel control and break it into three columns by placing three div tags with col-md-4 class. Each column will contain a card showing a picture and some text. To do this, take the HTML Card template from the Bootstrap documentation and place it in each column respectively. This will result in the following screen:

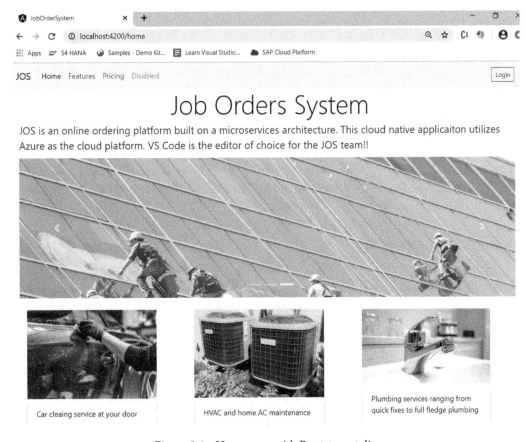

Figure 5.6 – Home page with Bootstrap styling

The preceding screenshot shows the final output—the home page for the job ordering system looks quite decent now. Let's move on and create our feature modules.

Creating the feature modules

As we discussed earlier, to separate concerns and organize our app into manageable blocks, Angular provides an option to break your application into modules. These modules also serve the purpose of loading the application more efficiently.

As explained in the application design, we will be dividing our app into three feature modules, as follows:

- `LoginModule`
- `JobrequestModule`
- `AgentModule`

Each of these modules will contain the routing paths specific to the functionality they encapsulate.

To create the module, we use `ng generate module <module_name>`, and also enable routing by adding the `--routing` option to the command line.

Creating the login feature

The login feature will provide the login page for the application. Our focus currently will only be on developing the login page. In a real-life scenario, you could have additional functionality such as signing up, or resetting a password/requesting help with a forgotten password.

Creating the login module

To create a feature, we start by creating a module. This module will hold the components for this feature. You can use the following command to generate `LoginModule` using the CLI:

```
ng generate module login --routing
```

The CLI will create a `login` folder and generate two files. The first file, `login.module.ts`, defines the `LoginModule`, and since we added the `routing` option, the CLI also created the `login-routing.module.ts` file. The `LoginRoutingModule` will contain the routing paths for `LoginModule`.

Here is the code for the generated `LoginRoutingModule`:

```
@NgModule({
    imports: [RouterModule.forChild(routes)],
    exports: [RouterModule]
})
export class LoginRoutingModule { }
```

The following code is for the generated `LoginModule`:

```
@NgModule({
    declarations: [],
    imports: [
        CommonModule,
        LoginRoutingModule
    ]
```

```
})
export class LoginModule { }
```

Since the route paths for the LoginModule will be defined in the
LoginRoutingModule, the CLI has already imported this in to the LoginModule.

Creating the login component

Next, let's generate the login component. Since we want this component to be part of the
login module and not the default app module, we will use the --module option.
Run the following command in the terminal to generate the LoginComponent:

```
ng generate component login --module=login
```

The CLI will generate the LoginComponent TypeScript, HTML, and CSS files. Since we
added the –module=login option, the login component was added to the declaration
section in the LoginModule and not the AppModule.

To update the login form in the login.component.html file, go to http://
getbootstrap.com, and then select **Documentation | Forms**. From here, copy the
Login Form and replace the existing code in the login.component.html file. Add a
div tag around the form and delete the form group containing the **Check Out** checkbox.
Replace the button text from Submit to Login and change the Bootstrap button styling
class from btn-primary to btn-dark, and place an h1 tag to give the login form
heading, as follows:

```
<h1 class='display-4'>Login</h1>
```

To submit the form, we listen to the ngSubmit event. As soon as the login button of type
Submit is pressed, we call the login method of the LoginComponent, like this:

```
<form (ngSubmit)='login(loginForm)' #loginForm='ngForm'>
```

Clicking the Submit button passes the loginForm object to the login method. Using
the #loginForm='ngForm' construct provides access to the form values in the login
method.

To pass the `username` and `password` fields with the `loginForm` object, define the `ngModel` directive with the two form fields, as follows:

```
<input type='email' id='email' name='userName' ngModel
class='form-control'>
<input type='password' id='password' name='password' ngModel
class='form-control'   >
```

The `login` method receives the form data and calls `AuthorizationService` for authentication and authorization. Upon a successful outcome, the user is redirected to the `Job Request` or `Agent` page; in the case of an error, a message is displayed. We will be covering the `AuthorizationService` later in this chapter.

The `NgModel` and `NgForm` directives are part of `FormsModule`. Import `FormsModule` in the `LoginModule`.

Configuring routing

Each feature module will specify the routing paths for its components. As of now, the `LoginModule` has only one component, so we set this as the default path for the `LoginModule`. The login feature can have more components, such as **Forgot Password** or **Reset Password**. Have a look at the following code snippet:

```
const routes: Routes = [
    { path:'', component: LoginComponent }
];
```

The preceding code snippet defines `LoginComponent` as the default path for the login feature.

Enabling lazy loading

As we discussed earlier, feature modules help in increasing the load performance of the app. This is done by using lazy loading.

An Angular app is bundled into a JavaScript file. Once the user hits the site URL, the server sends the same **JavaScript (JS)** file to the client. This will work fine for a smaller app, but in the case of an enterprise-grade application with several components, the app will take a considerable amount of time to load.

To increase performance, Angular provides a method to split the app, which can be loaded in chunks.

You must have noticed that we have already split our app into features, and each feature declares its components. To achieve lazy loading, instead of importing these modules into the `AppModule`, we will specify them under a parent path in the `AppRoutingModule`.

To hook up the `LoginModule` with the `AppModule`, add a routing path for the login feature module in the `AppRoutingModule`.

In the `app-routing.module.ts` file, we add the `login` parent path for all the paths inside our `LoginModule`. With the `loadChildren` parameter, we load the child routes of the `LoginModule`. The following code will tell Angular to load the `LoginModule` and its components only when the user routes to the login path. This way, when the user visits the site URL, only `AppModule` and its components are loaded, which drastically increases the initial load performance:

```
{
    path: 'login',
    loadChildren: () => import('./login/login.module').then(
                    module=>module.LoginModule)
}
```

Angular provides another neat option that can tell the application to asynchronously keep loading other feature modules while the landing page is already displayed in the browser, as illustrated in the following code snippet:

```
imports: [RouterModule.
forRoot(routes, { preloadingStrategy: PreloadAllModules})],
```

The preceding code snippet shows how to define the preloading strategy to asynchronously load the app.

Using the router link directive

To direct the user to the login form, we will add a login button to the navigation bar. Clicking this button will redirect the user to the login page. To add the login button in the `NavBar`, add the `routerLink` directive to the **Login** button we added earlier and specify the parent path `/login` for the `LoginModule`, as illustrated in the following code snippet:

```
<form class='form-inline'>
    <button class='btn btn-sm btn-outline-
```

```
secondary' type='button'
routerLink='/login'>Login</button>
  </form>
```

This will give you the following output:

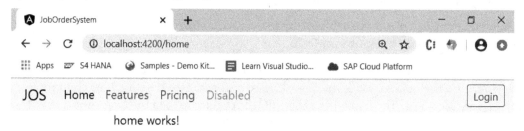

home works!

Figure 5.7 – Navbar with the Login button

The `routerLink` directive added to the **Login** button specifies the `/login` parent path for the `LoginModule`. On clicking the button, Angular will load the default route path of the `LoginModule` and replace the home page with the `LoginComponent` HTML template.

Creating the authorization service

Our login page is ready and linked to the home page. Let's develop an authorization service to handle authentication and authorization.

Implementing an identity provider is not part of the scope of this book, but to showcase route protection and login functionality, we will implement a simple authorization service, as follows:

1. Start by generating a service. To do this, run the following command in the terminal:

   ```
   ng generate service login/authorization
   ```

2. Next, generate the interface, as follows:

   ```
   ng generate interface models/loggeduser
   ```

3. In the generated interface, declare the following variables:

   ```
   export interface Loggeduser {
       userid: string;
       name: string;
   ```

```
      role: string;
   }
```

4. In the `AuthorizationService`, declare an object of type `LoggedUser`, like this:

```
user : Loggeduser;
```

The service will provide a `login` method to be called by the `LoginComponent`. The `AuthorizationService` will use the browser's `localStorage` to store the user object. The method checks `userid` and `password` against fixed values and returns a failure message in the case of a mismatch. In the case of success, the route path for the respective module is returned to the `LoginComponent`.

In the `LoginComponent`, we inject `AuthorizationService` and call its login method. Upon success, we use the `Router` service to redirect to the returned `redirectURL`, which will be the `Agent` or `User` page based on the role.

> **Tip**
>
> Please refer to the complete code in the Git repository here, `https://github.com/PacktPublishing/Visual-Studio-Code-Up-and-Running/tree/master/Chapter05`.

Switching Login and Logout buttons

To replace the **Login** button with **Logout** on the Navbar, the `AuthorizationService` provides the `isUserLoggedIn` method.

Inject the `AuthorizationService` by adding it in the `AppComponent` constructor, as follows:

```
constructor(public authorizationService: AuthorizationService)
{}
```

To hide and unhide the Login and Logout buttons, we put a `div` tag around them and include the `*ngIf` directive. The `*ngIf` directive calls the `isUserLoggedIn` method of the `AuthorizationService` to check whether the user session is active, as illustrated in the following code snippet:

```
<form class='form-inline'>
  <div *ngIf='!authorizationService.isUserLoggedIn()'>
<button type='button' routerLink='/login'>Login</button>
```

```
    </div>
    <div *ngIf='authorizationService.isUserLoggedIn()'>
  <button type='button' (click)='logOut()'>Logout</button>
    </div>
  </form>
```

We also add the `logout` method in our `AppComponent`. This method will call
the logout method of the `AuthorizationService`. The `logout` method of the
`AuthorizationService` resets the user object, as illustrated in the following code
snippet:

```
logOut():void{
    this.authorizationService.logout();
    this.router.navigate(['/home']);
}
```

Creating the job request feature

The job request feature provides the functionality to create and view job requests.
Following the same approach, we will generate the module, components, and the required
service.

Creating the job request service

If you recall, the job request feature module will use the Node.js backend to GET and
POST jobs from Azure Cosmos DB for the MongoDB **application programming
interface (API)** collection. To do this, we generate the `JobrequestService` using the
following CLI command. Create a `data_service` folder under `src/app` to keep all
your data services together, as illustrated in the following code snippet:

```
ng generate service data_service/jobrequest
```

The service will provide three methods: `saveJobRequest`, `getJobRequestList`,
and `getJobRequestById`. As the name suggests, these methods will be used to save
a job request, fetch the list of jobs created by the user, and fetch a particular job detail
respectively.

To make HTTP calls to the backend, we use the `HttpClient` service provided by
Angular. For this, import `HttpClientModule` in `AppModule`.

Currently, the backend services are running locally on our machine and we will be using `http://localhost:3001` to access them. Once these services are deployed, the URL(s) will change, and we don't want to change them in several places in our app. To keep the backend URL(s) in one place, we create `ConfigService`, as follows:

```
ng generate service data_service/config
```

In `ConfigService`, add the Node.js URL, like this:

```
nodeUrl : string = 'http://localhost:3001';
```

To type the job request object, generate the `IJobrequest` interface, as follows:

```
ng generate interface models/jobrequest
```

Add the following attributes to the interface:

```
export interface IJobrequest {
    _id: string,
    jobType: string,
    jobDescription: string,
    requestDate: string,
    contactNo: string,
    address: string,
    city: string,
    status: string,
    requestedBy: string
}
```

Going back to the `JobrequestService`, we now inject the `HttpClient` service to make HTTP calls, `ConfigService` to specify the Node.js URL, and `AuthorizationService` to fetch the logged-in user credentials in the `JobrequestService` constructor. In an actual scenario, the `AuthorizationService` will be used to fetch the bearer authorization token for authorized access to the backend. We will not be covering this here, but the rest of the process just described is illustrated in the following code snippet:

```
constructor(private authorizationService: AuthorizationService,
            private http: HttpClient,
            private config: ConfigService) { }
```

Since calls to the HTTP backend services are asynchronous, the saveJobRequest method returns Observable to the caller. We use the post method of the HttpClient service to make POST requests. It passes three parameters: the service URL, the request object in the body, and the content type in the options parameter in the HTTP request header, as follows:

```
saveJobRequest(request: IJobrequest) : Observable<any>{
  return this.http.post(this.config.nodeUrl + '/jobs',request,
                        this.getOptions());
}
```

The getJobRequestList method makes a GET request using the HttpClient service and passes in the service URL. The method returns an Observable of type IJobrequest, as illustrated in the following code snippet:

```
getJobRequestList(): Observable<IJobrequest[]>{
    //Get the Logged in User
    return this.http.get<IJobrequest[]>(this.config.
nodeUrl + '/jobs');

    }
```

The getJobRequestById method makes a parametrized GET call to the backend by passing in requestId, as illustrated in the following code snippet:

```
getJobRequestById(requestId: string): Observable<any>{
    return this.http.get<IJobrequest>(this.config.nodeUrl + '/
job?id='+requestId);
    }
```

With this, we have completed the job request service and exposed the required methods to access the backend Node.js API. Next, let's work on generating the components.

Creating the job request module

As we discussed earlier, each feature will be divided into an Angular module. To develop the job request feature components, we will start by generating the job request module. To do this, use the following CLI command:

```
ng generate module --routing jobrequest
```

Creating the job request list component

The JobrequestListComponent module will list the jobs created by the user; with the --module addition, we tell the CLI to declare this component in JobrequestModule, as follows:

```
ng generate component jobrequest/jobrequestlist
--module=jobrequest
```

The preceding CLI command will create the job request list component and its HTML and CSS file in the jobrequestlist folder under the jobrequest folder.

JobrequestModule will specify the routes for the job request feature. Let's add the JobrequestlistComponent component as the default route path for this module, as follows:

```
{ path: '', component: JobrequestlistComponent},
```

To lazy load the feature routes in the AppModule, add the Job request feature parent path in the AppRoutingModule, as follows:

```
{
    path: 'jobrequest',
    loadChildren:() => import('./jobrequest/jobrequest.
module').then(
                        module=>module.JobrequestModule)
}
```

If you recall, the `LoginComponent` will already redirect the user to the respective feature module. Additionally, we can add the link in the `NavBar`, as follows:

```
<li class='nav-item'>
  <a class='nav-link' routerLink='/
  jobrequest'>Job Request</a>
</li>
```

To fetch a list of created jobs from the Node.js backend, `JobrequestlistComponent` will call the `getJobRequestList` method of the `JobrequestService`. For this, we inject the `JobrequestService` in to the component constructor, as follows:

```
constructor(private service: JobrequestService) { }
```

The `getJobList` method of the component calls the service method and subscribes to an `Observable`. An `Observable` can return three responses: a first function in the case of success; a second function in the case of an error; and a third function once the request is completed. We will use the first two functions. A successful response is stored in the `requestList` array of type `IJobrequest`. In the case of failure, the `handleError` method displays an error to the user. Please refer to the complete code from the Git repository, a snippet of which is shown here:

```
getJobList(){
  this.service.getJobRequestList().subscribe(
    res=> { this.requestList = res },
    error=> this.handleError(error)
  );
}
```

To populate the list as soon as the **Job Request** page is displayed, we call the `getJobList` method in the `ngOnInit` Angular lifecycle hook. This method is called during the initialization process, and is illustrated in the following code snippet:

```
ngOnInit(): void {
  this.getJobList();
}
```

The list component is ready, so let's update its HTML template in `jobrequestlist.component.html`.

We will be using a table to show a list of jobs. You can visit the `https://getbootstrap.com/docs/4.1/content/tables/` Bootstrap link and look for a suitable table of your choice.

To create our table using the **Emmet** snippet, we write the following command in the HTML template. Use *Tab* to jump to each column:

```
table.table>thead>tr>th*4
```

Inside the `tbody` tag, write the following command to create a row with four columns:

```
tr>td*4
```

To loop through the list of jobs and display them in a table, we use the `*ngFor` directive. In the following code snippet, we declare a runtime request object and loop on the `requestList` array. Using interpolation, display the value of the request object in the `td` tags:

```
<tbody>
  <tr *ngFor='let request of requestList'>
    <td>{{request.jobType}}</td>
    <td>{{request.jobDescription}}</td>
    <td>{{request.requestDate}}</td>
    <td>{{request.status}}</td>
  </tr>
</tbody>
```

To show an alert in the case of an error, we use the `*ngIf` directive and check whether `showError` is set to `true`:

```
<div *ngIf='showError' class='alert alert-danger'>
    {{errorMessage}}
</div>
```

The `showError` variable is set to true if the `JobrequestService` method returns an error. In the `handleError` method of the `JobrequestlistComponent`, we set `showError` to true and also set `errorMessage` for the user.

Creating a job request form

The `JobrequestComponent` will provide a form to create new job requests. Use the following command to generate the component, and update the `JobrequestModule`:

```
ng generate component jobrequest/jobrequest --module=jobrequest
```

Add the routing path in `JobrequestRoutingModule`, as follows:

```
{ path: 'request', component: JobrequestComponent }
```

After login, the user is redirected to the **Job Request List** page. Here, we will provide a button to create a new job request. The button will route to the `JobrequestComponent` by specifying the `routerLink` directive. The code for this can be seen in the following snippet:

```
<button type='button' routerLink='/jobrequest/request'>
Create Job Request</button>
```

The preceding code snippet calls the `/request` child route of the `/jobrequest` job request feature module parent route.

Next, we move on and define the HTML template for our component. Since we will be creating a form, the first thing to do is to import the `FormsModule` in the `JobrequestModule`.

Update the HTML template in the `jobrequestnew.component.html` file. We will have a dropdown with the type of job; a text area where the user can provide a description of the problem; a preferred date and time; **Contact No**, **Address**, and **City**.

Exploring two-way data binding

In our previous example of the **Login** form, we used the `ngForm` object to access form values.

Here, we will explore Angular's two-way binding.

For two-way binding to work, we will first create an object of the `IJobrequest` interface in `JobrequestComponent`, as follows:

```
jobRequest = {} as IJobrequest;
```

We will link the request object to each form field using `[(ngModel)]`. In the following code snippet, we bind the `requestDate` parameter of the `jobRequest` object to the form input element, and `jobRequest` is declared in the `JobrequestComponent`:

```
<input id='requestDate' class='form-control' type='datetime-
local' name='requestDate' [(ngModel)]='jobRequest.requestDate'>
```

Similarly, you can add the remaining fields of the form in the HTML template. Please refer to the code in the Git repository.

Creating form validations

We would like our users to enter the right information when submitting their requests. For this reason, it's important to implement form validations.

Angular provides specific classes to manage these validations—these are `ng-untouched`, `ng-touched`, `ng-pristine`, `ng-dirty`, `ng-valid`, and `ng-invalid`.

You can check their description at `https://angular.io`.

For each form control, these properties can be accessed by referencing `ngModel`. In the following example, we access `ngModel` by assigning it to `contactProperty`:

```
<input id='contactNo' required class='form-control'
type='tel' #contactProperty='ngModel' name='contactNo'
[(ngModel)]='jobRequest.contactNo'>
```

With the `contactProperty` variable, we can now use the ng properties mentioned previously and display conditional help for the user.

The following code example unhides the span once the user clicks and removes the cursor from the field, or empties it after entering a value. We can further display specific messages for each type of error, such as required, min, max, and so on. Here, we show a message for the required error:

```
<span class='help-block'
 *ngIf='(contactProperty.touched || contactProperty.dirty)
  && contactProperty.errors'>
  <small *ngIf='contactProperty.errors.
required'>Please enter contact no. </small>
</span>
```

Following the same approach, you can place validations for each field and put specific messages.

> **Tip**
>
> If you recall from *Chapter 1, Getting Started with Visual Studio Code* you can highlight contactProperty and use *Ctrl + D* to quickly select and replace the variable name using VS Code multi-cursor editing.

Finally, to make sure that the user can only submit the request once the form is filled out correctly, we use the ngForm properties to disable the button until it reaches a valid state.

In the form tag, we add a requestForm variable to access ngForm, as follows:

```
<form #requestForm='ngForm' (ngSubmit)='onSubmit(requestForm)'>
```

We use it to disable the **submit** button. The disabled property will be set to true if the requestForm.valid property is false. The code can be seen in the following snippet:

```
<button type='submit' [disabled]='!requestForm.valid'
class='btn btn-dark'>Send Request</button>
```

Our **Create Job Request** form is complete—here is how it looks:

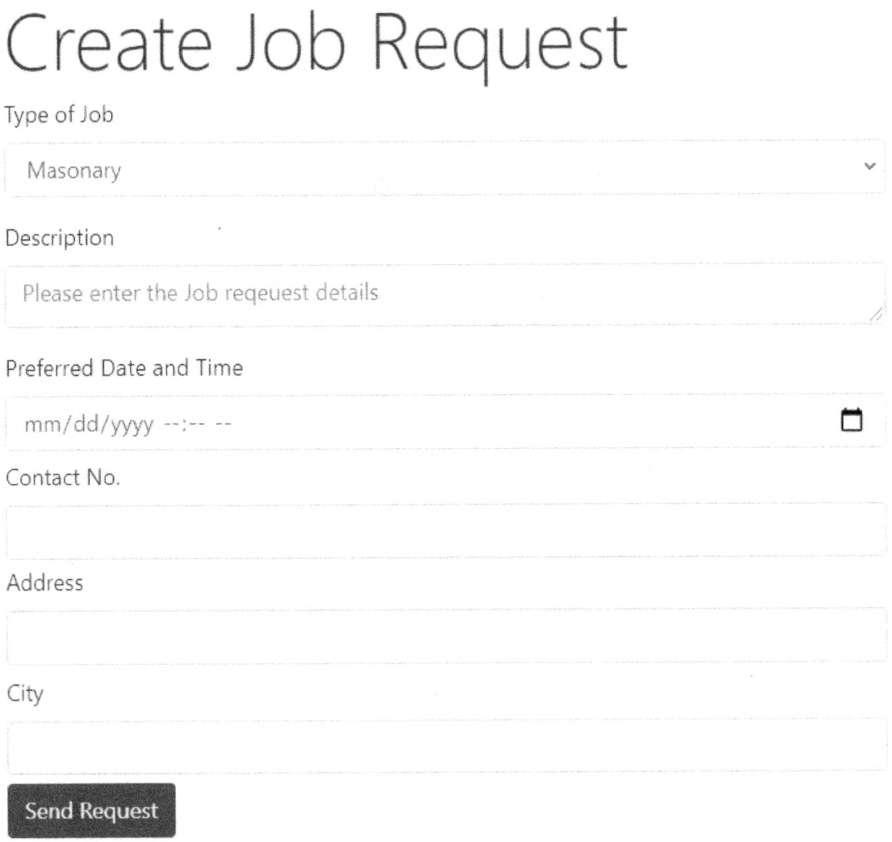

Create Job Request

Type of Job

| Masonary | ⌄ |

Description

| Please enter the Job reqeuest details | |

Preferred Date and Time

| mm/dd/yyyy --:-- -- | 🗓 |

Contact No.

| |

Address

| |

City

| |

Send Request

Figure 5.8 – Create Job Request form

Next, let's move on and link this form to the backend Node.js API.

Earlier in this chapter, we created `JobrequestService`. The service provides the `saveJobRequest` method to post new requests in the Node.js backend.

We start by injecting the `JobrequestService` in to the `JobrequestComponent` by declaring it in the constructor, as follows:

```
constructor(private service: JobrequestService) { }
```

Next, call the `saveJobrequest` method of the `JobrequestService` in the `onSubmit` method. As shown previously, the `onSubmit` method is linked to the `ngSubmit` event of the form.

In the `onSubmit` method, we call the `saveJobrequest` method and subscribe to its **Observable**.

The `handleSuccess` method will be called when the HTTP service returns successfully, and the `handleError` method will be called in the case of an error being returned from the Node.js backend, as illustrated in the following code snippet:

```
onSubmit(form: NgForm){
    this.service.saveJobRequest(this.jobRequest).subscribe(
        res => this.handleSuccess(res),
        error => this.handleError(error)
    );
}
```

The `handleSuccess` and `handleError` methods will show success and error messages to the user respectively, as illustrated in the following code snippet:

```
handleSuccess(res){
    this.showSuccess = true;
    this.successMessage = res;
}

handleError(error){
    this.showError = true;
    this.errorMessage = error;
}
```

These variables are declared in the `JobrequestComponent`, as follows:

```
showError: boolean = false;
errorMessage: string;
showSuccess: boolean = false;
successMessage: string;
```

It's very important for an application to communicate messages in an intuitive manner. For this, let's use Bootstrap classes to display the success and error messages.

Displaying user messages

To show return messages from the server, we add two `div` tags in our `jobrequest.component.html` file. The respective `div` tag is shown when either `showSuccess` or `showError` is set to true by the `handleSuccess` or `handleError` methods respectively. We use the Bootstrap alert classes to style them for success and errors respectively, as illustrated in the following code snippet:

```
<div *ngIf='showSuccess' class='alert alert-success'>
    {{successMessage}}
</div>
<div *ngIf='showError' class='alert alert-danger'>
    {{errorMessage}}
</div>
```

In both the `div` tags mentioned previously, we use the Angular `*ngIf` construct to show only the relevant message.

Creating a job request detail page

The last thing before we complete our job request feature is to provide the user with an option to view the details of each job request they created. We will show a button in each table row and upon this being clicked, open the details of the selected job. To do this, proceed as follows:

1. Start by generating `JobrequestdetailComponent`, as follows:

    ```
    ng generate component jobrequest/jobrequestdetail
    ```

2. Next, we add a routing path in the `JobRequestRoutingModule`. This will take the job request ID as a routing parameter. The code can be seen in the following snippet:

    ```
    { path: 'request/:id', component:
    JobrequestdetailComponent }
    ```

3. To call the route, we add a `Display` button in our table with a `routerLink` directive. Since the route parameter will be different for each row, we enclose the `routerLink` directive in square brackets and also pass the `requestId` parameter as the second object in the array.

4. The `/jobrequest/request` path tells Angular to call the `request` route path of the `jobrequest` parent path. If you recall, the `jobrequest` parent path is already defined in the `AppRoutingModule`, and the `request` path is its child defined in `JobRequestRoutingModule`. This is illustrated in the following code snippet:

```
<td>
        <button type='button'
        [routerLink]='['/jobrequest/request',
request.requestId]'>Display
        </button>
</td>
```

5. Inject the `JobrequestService` in the component constructor, as follows:

```
constructor(private service:JobrequestService) { }
```

6. Add a `jobRequest` object to store the request details, as follows:

```
jobRequest = {} as IJobrequest;
```

7. The following code defines a component method to call the `getJobRequestById` method of the `JobrequestService`, to fetch job details. In the case of success, a response is assigned to the `jobRequest` object, as illustrated in the following code snippet:

```
getJobById(jobId: string) {
    this.service.getJobRequestById(jobId).subscribe(
        res => { this.jobRequest = res; },
        error => { return this.handleError(error); }
    )
}
```

The `handleError` method is similar to our `JobrequestComponent`.

Reading the route parameters

To read the route parameters, we will use the Angular `ActivatedRoute` service. Inject this service in to the `JobrequestdetailComponent` constructor, as follows:

```
constructor(private service: JobrequestService,
            private activatedRoute: ActivatedRoute) { }
```

Using the activatedRoute object, we fetch the router parameters passed by the JobrequestlistComponent and call the getJobById method defined previously to fetch the job details. To fetch the job details on load, we use the ngOnInit lifecycle hook, as follows:

```
ngOnInit(): void {
    const requestId = this.activatedRoute.snapshot.paramMap.
get('id');
    this.getJobById(requestId);
}
```

> **Note**
>
> The ActivatedRoute service also provides an observable through which you can listen to changes in the URL parameter while staying in the same component, as follows:
>
> ```
> this.activatedRoute.paramMap.subscribe(
>     params => { const requestId = params.
> get('id'); }
>     )
> ```

Finally, we update the jobrequestdetail.component.html template to display the job details. Please refer to the Git repository for the complete code.

Creating the agent feature

The agent feature will provide the agent with the functionality to view and respond to assigned jobs. Similar to the job request feature, follow these next steps:

1. Generate the service and the interface, as follows:

    ```
    ng generate service data_service/agent
    ```

2. Generate the interface, like this:

    ```
    ng generate interface models/agent
    ```

3. Inject the AuthorizationService and the HttpClient service in the constructor and create the saveAgentAction, getAgentJobList, and getAgentJobById methods.

4. Since the agent backend services are built using a separate microservice on the Java Spring Boot framework, we will add another URL to the `ConfigService` and refer to that in the service methods, as follows:

```
javaUrl : string = 'http://localhost:9003';
```

Creating the agent module

Following a similar approach, we generate the agent module, as follows:

```
ng generate module --routing agent
```

The preceding command will create the agent and agent routing module. Next, we create the agent module components.

Creating the agent list component

The agent list page will display the list of jobs assigned to an agent, as follows:

1. Start by generating the agent list component, as follows:

```
ng generate component agent/agentlist --module=agent
```

2. First, add the routing path to `AgentRoutingModule`, as follows:

```
{ path:'', component: AgentlistComponent },
```

3. Hook `AgentModule` to `AppModule`, specifying the agent route path in the `AppRoutingModule`, as follows:

```
{
    path: 'agent',
    loadChildren:() => import('./agent/agent.module').then(
                        module=>module.AgentModule)
}
```

4. Declare an array of `IAgent` interface to hold the list of jobs pending with the logged-in `Agent`.

5. Inject the `AgentService` in to the constructor.

6. Create the `getAgentJobs` method, which should call the `getAgentJobList` method of the `AgentService`.

7. Call the `getAgentJobs` method in the `ngOnInit` lifecycle hook.

8. Update the `agentlist.component.html` template.

Creating the job processing component

The job processing component displays the job details to the agent. It also allows the agent to update the job status and insert remarks as follows:

1. Generate the `AgentJobprocess` component, as follows:

    ```
    ng generate component agent/agent-jobprocess
    --module=agent
    ```

2. Add a parameterized route in `AgentRoutingModule`, as follows:

    ```
    { path:'request/:id', component:
    AgentJobprocessComponent}
    ```

3. Declare an object of type `IAgent` in `AgentJobprocessComponent`. This will hold the form data. The code can be seen here:

    ```
    agentJob = {} as IAgent;
    ```

4. Inject the `AgentService` and `ActivatedRoute` in to the component constructor, as follows:

    ```
    constructor(private service:AgentService,
                private activatedRoute:ActivatedRoute) { }
    ```

5. Similar to the `JobrequestdetailComponent`, create the `getAgentJobById` method in `AgentJobprocessComponent`.

6. Read the route `param` ID and call the `getAgentJobById` method in the `ngOnInit` lifecycle hook, as follows:

    ```
    ngOnInit(): void {
      const requestId = this.activatedRoute.snapshot.
    paramMap.get('id');
      this.getAgentJobById(requestId);
    }
    ```

7. Create the `processAgentJob` method and call the `saveAgentAction` method of `AgentService`, as follows:

```
processAgentJob(form: NgForm) {
    this.service.saveAgentAction(this.agentJob).
    subscribe(
      res => this.handleSuccess(res),
      error => this.handleError(error)
    );
}
```

8. Update the `agent-jobprocess.component.html` template and create a form where status and agent remarks fields are editable.

Processing routes with authorization guard

Our application is complete and the routing works as expected. But there is a small problem. If you try to hit the `http://localhost:4200/agent` or `http://localhost:4200/jobrequest` URL without logging in, the app will redirect you to the respective page. This is happening since our routes are not protected.

To protect our routes, Angular has provided the `AuthGuard` interfaces. We will implement them to protect unauthorized access to our application.

You can get more details about different types of route guards from the Angular website. Here, we will implement the `CanActivate` route guard.

To check user authorization before the route is activated, we will use the `CanActivate` guard.

We will generate two guards: one for our job request feature and another one for our agent feature. We can generate guards to restrict access for each route, but since our requirement is to allow user access for all job request features and agent access to all agent features, we will create and assign guards for the job request and agent paths and not for each child route path.

Creating a job request guard

Use the following CLI command to generate the guard for the job request feature:

```
ng generate guard jobrequest/jobrequest
```

This command will give you the following prompt:

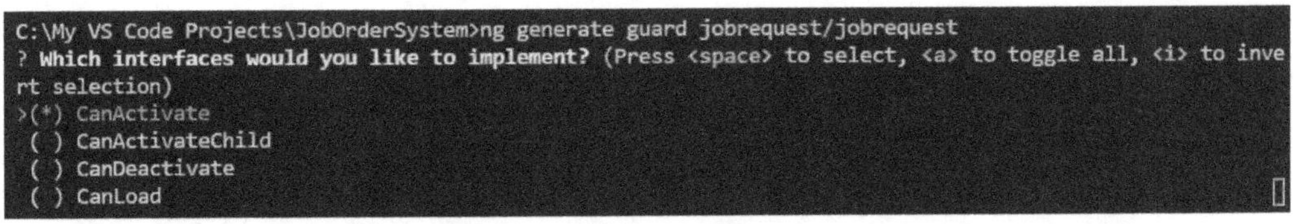

```
C:\My VS Code Projects\JobOrderSystem>ng generate guard jobrequest/jobrequest
? Which interfaces would you like to implement? (Press <space> to select, <a> to toggle all, <i> to inve
rt selection)
>(*) CanActivate
 ( ) CanActivateChild
 ( ) CanDeactivate
 ( ) CanLoad
```

Figure 5.9 – Shows the CLI prompt for selecting the type of guard to be generated

The CLI will prompt you to select the specific guard you would like to generate. By default, CanActivate is selected with an *. Press *Enter* and continue. The guard is generated. Next, we will implement the canActivate method of the CanActivate interface.

Our objective is to allow access to the route if the user is logged in. Additionally, since our application has two roles—one for the user, who can access the job request feature, and another for the agent, who can access the agent feature—we will also check the role property. If any of the checks fail, we will redirect the user to the login page.

First, inject the AuthorizationService and the Router service in to the guard service constructor, as follows:

```
constructor(private authorizationService: AuthorizationService,
private router:Router) { }
```

In the canActivate method, we check the user login status by calling the isUserLoggedIn method of the AuthorizationService. Secondly, we fetch the requested route path by accessing the ActivateRouteSnapshot object passed to the canActivate method. The URL path is passed as an array. We access the first object of the array and compare it with the logged-in user role in the AuthorizationService. For the jobrequest route path, the role should be user. If all conditions are met, the method returns true, and the app will proceed to the job request list page. If any of the preceding conditions fail, we route to the login page. The code for this can be seen here:

```
canActivate(
    next: ActivatedRouteSnapshot,
    state: RouterStateSnapshot): Observable<boolean |
UrlTree> | Promise<boolean | UrlTree> | boolean | UrlTree {

    let url = next.url[0].path;
    if (this.authorizationService.isUserLoggedIn()) {

    if (url == 'jobrequest' &&
        this.authorizationService.user.role == 'user')
```

```
        return true;
    else
        return false;
    }
    else{
        this.router.navigate(['/login']);
    }
}
```

Finally, we specify JobrequestGuard in AppRoutingModule to protect the jobrequest route path, as follows:

```
{
    path: 'jobrequest',
    loadChildren:() => import('./jobrequest/jobrequest.module')
.then(
                            module=>module.JobrequestModule),
    canActivate:[JobrequestGuard]
},
```

The preceding addition in the routing path will call the JobrequestGuard and check whether it's allowed to proceed and load the component. Our job request module is now protected, so let's follow the same approach to protect the agent feature module as well.

Creating the agent guard

The agent guard will be used to protect the route paths for the agent feature module.

Follow the same steps to generate and code AgentGuard, as follows:

1. Generate the guard, as follows:

    ```
    ng generate guard agent/agent
    ```

2. Implement the canActivate method.

3. Specify the AgentGuard in the AppRoutingModule to protect the agent route path.

Summary

In this chapter, we completed the frontend of our job order system. We started by giving an introduction to Angular and moved on to discuss the application design. Before starting the development, we set up the environment and installed the Angular CLI.

During this chapter, we learned about how an enterprise-grade application can be developed using Angular in VS Code. We learned about the Angular CLI and how it can be used to speed up the development process. We also learned advanced techniques to enhance app performance and implement security. For styling, we learned how to use the Bootstrap toolkit and enhance the appearance of our application.

In the next chapter, we will look into debugging techniques and explore debugging extensions for the different platforms we have used. We will explore navigation, breakpoints, logpoints, and data inspection, and will also learn about advanced debugging concepts.

6
Debugging Techniques

In *Chapter 5, Building a Web-Based Frontend Application with Angular* we completed the frontend application and, with this, the job order system case study has been completed. The focus of this book has been on developing a multiplatform application, based on a microservices architecture. We started the book with an introduction to **Visual Studio Code (VS Code)** features, and then moved on to developing the services and the frontend. If you have been following along, or have already downloaded the code base from the Git repository, we will now use the code base and explore the debugging features of VS Code.

Writing code is one part of programming, and efficiently debugging issues and solving them is another important part. People follow different techniques to locate bugs and fix them. This varies from writing console logs to using the extensive features provided by the development tool. Ultimately, you will find and fix the issue, but the real question is about how much time it takes.

Our focus in this chapter will be to explore how we can debug the different code bases. We will look at the different extensions available and how they can be set up in VS Code. VS Code falls between the editor and **integrated development environment** (**IDE**) space, and that is predominantly because of the extended tooling features it provides.

By the end of this chapter, you will have learned about the following topics:

- Becoming familiar with the debugger layout
- Debugging the Angular app
- Debugging features of VS Code
- Debugging the Node.js **application programming interface** (**API**)
- Debugging the Java API
- Debugging the Python API
- Debugging .NET Core

We will also explore breakpoints, variables, the call stack, and logpoints while debugging the application. Further, we will see how to create launch configurations.

So, let's get started and explore the extensive features of VS Code and how extensions help expand them even further.

Becoming familiar with the debugger layout

First things first: before we start installing extensions and configuring the debugger, let's look at the layout of the debugger. Once you get accustomed to the layout, it will be easier to navigate. To call the debugger, press the button on the left bar, or press *Ctrl + Shift + D* to reveal the following screen:

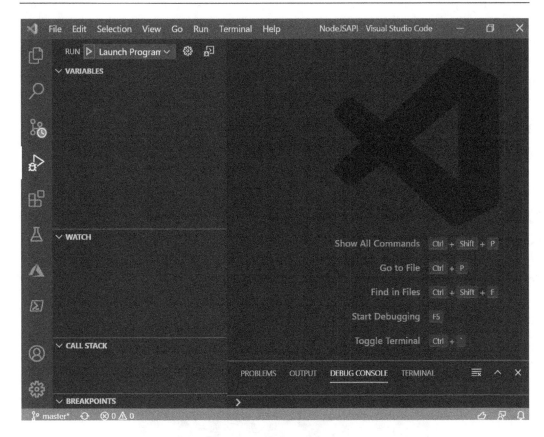

Figure 6.1 – Debugger layout

The top bar in the left pane holds the **RUN** button, as shown in the preceding screenshot. This is used to launch the debugger. VS Code uses the launch configuration to configure the debugger for debugging different code bases. You can select the respective configuration from the dropdown beside it and click **RUN**. The keyboard shortcut for the **RUN** command is *F5*. The configuration toolbar can be seen in the following screenshot:

Figure 6.2 – Configuration toolbar

The **Settings** button in *Figure 6.2* opens the selected launch file in the right pane. While debugging, you will see a list of global and local **VARIABLES**, **CALL STACK**, or **BREAKPOINTS** in their respective sections, as shown in *Figure 6.1*. While debugging, you can also add watch points from the **WATCH** section.

The last button in the top bar is for the debug console. Clicking it will open the debug console in the bottom pane.

With an introduction to the debugger and its different sections, let's start by debugging the Angular app.

Debugging the Angular app

Extensions are one of the most important features of VS Code, and we have explored this utility several times in the previous chapters. VS Code is supported by an extensive list of debugging extensions, which we will see in the following subsections.

Installing the debugger for Chrome

The first extension we will look at is the **Debugger for Chrome** extension. Let's install it:

1. Go ahead and install it from the **EXTENSIONS** tab, as illustrated in the following screenshot:

Figure 6.3 – Debugger for Chrome extension

2. Once the extension is installed, go to the **Debugger** tab, or press *Ctrl + Shift + D*. As shown in *Figure 6.4*, VS Code will ask you to create a launch.json file. This launch file contains the configuration for the debugger, and the message to create it can be seen here:

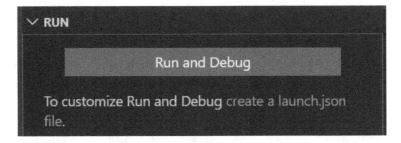

Figure 6.4 – Option to create a debugger launch.json file

3. On clicking **create a launch.json file**, the **Command Palette** will open and ask for the environment. Select **Chrome**, as shown in the following screenshot:

Figure 6.5 – Select the debugging environment

VS Code will create a `.vscode` folder in your project directory and will generate the launch configuration.

Discussing the launch configuration

The launch configuration is the basis for debugging any code base in VS Code. There are a couple of ways to launch the debugger. You can either *launch* a new browser or *attach* it to an existing session.

Launch request

The `launch` request type will start a new browser session and link the debugger. The following **JavaScript Object Notation (JSON)** code shows how the configuration for a `launch` request will look:

```
'configurations': [
    {
        'type': 'pwa-chrome',
        'request': 'launch',
        'name': 'Launch Chrome against localhost',
```

```
            'url': 'http://localhost:4200',
            'webRoot': '${workspaceFolder}'
    }
```

The configuration starts with the `type` as `pwa-chrome`. As we discussed, `request` can either be `launch` or `attach`; here, we use `launch`. The `launch` option will start a new browser window and load the debugger in VS Code. The `attach` option will attach it to the site loaded in the browser. The `url` parameter specifies the host and the port where the app is running, and `webRoot` is the project directory.

The Attach request

Another option that VS Code provides is the ability to attach the debugger to an existing browser session. The following JSON code configures the `attach` option:

```
    {
            'name': 'Attach to Chrome',
            'port': 9222,
            'request': 'attach',
            'type': 'pwa-chrome',
            'webRoot': '${workspaceFolder}'
    }
```

The `request` type is `attach`. In the case of `attach`, we will need to specify the `port` on which the browser will listen to the debugger. The remote debugging port can be specified by launching Chrome with the following command. Run this command in the directory where Chrome is installed:

```
chrome --remote-debugging-port=9222
```

On running this command, Chrome will open and start listening on port `9222` for any debug requests.

Running the debugger

Before you launch the debugger, open the terminal window, and run the Angular application using `ng serve`. To run the debugger, press *F5* or hit the **RUN** button in the DEBUG tab.

On successfully launching the debugger, VS Code will change the color of the bottom bar to red. In the debug tab, VS Code will start displaying the active breakpoints, call stack, loaded scripts, and so on. Let's look at these different features in more detail.

Debugging features of VS Code

We have set up VS Code for debugging our Angular frontend. Since we will require a code base to debug, here we will use the Angular app to explore different debugging features of VS Code. However, these features are common to other languages and can be used for debugging other code bases as well.

Exploring breakpoints

One of the most common techniques for debugging code is by placing breakpoints. **Breakpoints** tell the application to stop and transfer control to the debugger, from where the developer can execute the code line by line and check for errors. Let's look at how we can use breakpoints in VS Code.

Creating a breakpoint

Setting a breakpoint is pretty simple—click beside the line number where you want the debugger to stop, as illustrated in the following screenshot:

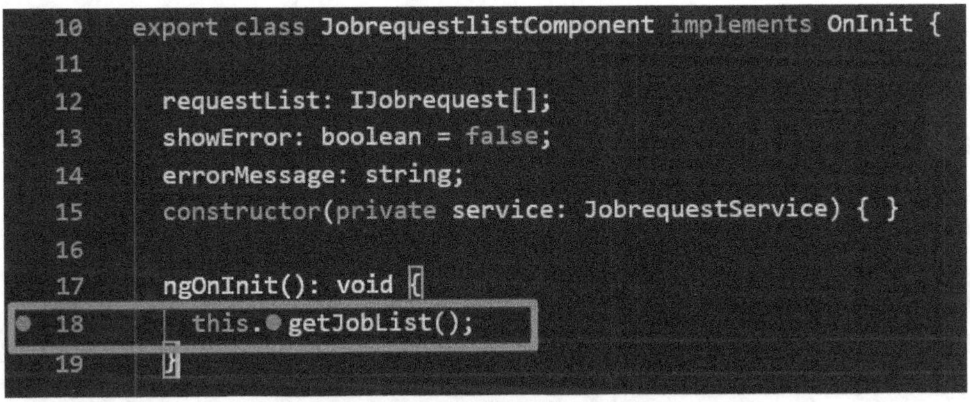

```
10      export class JobrequestlistComponent implements OnInit {
11
12          requestList: IJobrequest[];
13          showError: boolean = false;
14          errorMessage: string;
15          constructor(private service: JobrequestService) { }
16
17          ngOnInit(): void {
18              this. getJobList();
19          }
```

Figure 6.6 – Breakpoint set on line number 18

Figure 6.6 shows a breakpoint set in the ngOnInit() method of JobrequestlistComponent.

Calling a breakpoint

To call a breakpoint that we set previously, let's visit the **Job Requests** list page. `JobrequestlistComponent` fetches a list of jobs on initialization, using the `ngOnInit` lifecycle hook. While the page is being loaded, our set breakpoint will be triggered, and control will transfer back to VS Code, as shown in the following screenshot:

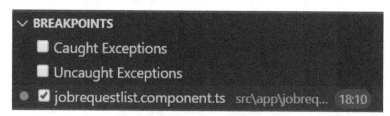

Figure 6.7 – Breakpoint called in VS Code

Once the debugger is called, in the left panel you will see a list of breakpoints, as illustrated in the following screenshot:

Figure 6.8 – List of breakpoints

The ones marked with a tick are active. The debugger also displays the filename and the line number where the breakpoint is set.

Deactivating a breakpoint

To deactivate a breakpoint, simply deselect the checkbox. This will disable the breakpoint, as shown in the following screenshot:

Figure 6.9 – Disabled breakpoint

VS Code will keep track of disabled breakpoints, but they will not be called while debugging. By selecting a breakpoint again, you can enable it.

Removing a breakpoint

To remove a breakpoint, you can click the red dot again or press *F9*, as illustrated in the following screenshot:

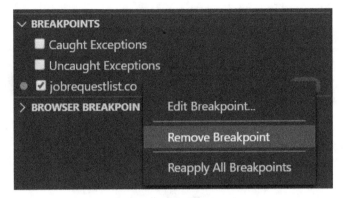

Figure 6.10 – Removing a breakpoint

Alternatively, you can also use the **Remove Breakpoint** menu in the **BREAKPOINTS** section, as shown in *Figure 6.10*.

Removing or deactivating all

VS Code also provides an option to quickly deactivate or remove all breakpoints. These buttons are available in the top bar of the **BREAKPOINTS** section, as shown in the following screenshot:

Figure 6.11 – Deactivate all and remove all respectively

You can use the following icon to deactivate a breakpoint:

Figure 6.12 – The deactivate icon

The following icon helps to remove data:

Figure 6.13 – The remove all icon

Conditional breakpoints

To stop the debugger based on a condition, you can use conditional breakpoints. To create a conditional breakpoint, right-click beside the line number and select **Add Conditional Breakpoint…**, as illustrated in the following screenshot:

Figure 6.14 – Add Conditional Breakpoint

Figure 6.14 shows you how to set a conditional breakpoint. Compared to normal breakpoints, these are represented in VS Code by a circle with two lines, as illustrated here:

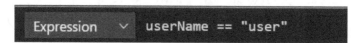

Figure 6.15 – Circle with two lines

To specify the condition on which the debugger should be called, write the expression as shown in the following screenshot:

Figure 6.16 – Expression for a conditional breakpoint

As soon as the condition mentioned in *Figure 6.16* is `true`, the debugger will take control and activate for debugging. In this expression variable, `userName` is compared with a `user` string, as illustrated in the following screenshot:

Figure 6.17 – Debugger stopping at a conditional breakpoint

When you enter `user` as the `username`, the debugger will break execution and stop for debugging, as shown in *Figure 6.17*.

After discussing and exploring breakpoints in detail, let's explore the different navigation options provided by the VS Code debugger.

Exploring navigation features

As with other IDEs, VS Code also provides navigation features. In this section, we will discuss how to navigate when a breakpoint is hit and the debugger is called.

The navigation bar is shown in the following screenshot:

Figure 6.18 – Navigation bar

You have already set the breakpoint, and the debugger has stopped execution for you to inspect and debug your code. The first thing you will notice is the navigation bar at the top of the window, as shown in *Figure 6.18*. Let's discuss how we can utilize its different functions, as follows:

- **Continue**: The first button, which is similar to a **play** button, is used to take the debugger to the next breakpoint in the code. If no further breakpoints exist, control will pass back to the web page. The keyboard shortcut for **Continue** is *F5*, and the button can be seen here:

Figure 6.19 – The Continue button

- **Step Over**: The next button is the **Step Over** button, or shortcut key *F10*. If you come across a method or a function while debugging, you can step over to navigate through without going inside the function code, as illustrated in the following screenshot:

Figure 6.20 – Stepping over a method call

The example in *Figure 6.20* stops the debugger on the getJobList method. Pressing *F10* or **Step Over** will pass the debugger to the next statement after the method call, which in our case is the curly brace.

- **Step Into**: **Step Into**, or shortcut key *F11*, is the opposite of **Step Over**. It will tell the debugger to navigate inside the method call, as illustrated in the following screenshot:

```
17      ngOnInit(): void {
18         this. getJobList();
19      }
20
21      getJobList(){
22         this.service.getJobRequestList().subscribe(
23            res=> { this.requestList = res; },
24            error=> this.handleError(error)
25         );
26      }
```

Figure 6.21 – Stepping into a method call

Referring to the same example in *Figure 6.20*, **Step Into** will navigate inside the getJobList () method, as shown in *Figure 6.21*.

- **Step Out**: **Step Out**, or shortcut key *Shift + F11*, is used to pass the cursor out of the called function or method. Suppose that you have navigated into a method, and you would like to go out to the next statement after the function call. You can use the **Step Out** function for this, as illustrated in the following screenshot:

```
17      ngOnInit(): void {
18         this. getJobList();
19      }
```

Figure 6.22 – Step Out

While debugging the getJobList method in *Figure 6.20*, using **Step Out** will tell the debugger to jump out of the current method and stop at the next line after the method call. *Figure 6.22* shows the result of using **Step Out** from within the getJobList method.

- **Restart**: To restart, click the **Restart** button, which is illustrated here:

Figure 6.23 – Restart button

- **Disconnect or stop**: We can use the disconnect or stop function to end the debugging session. When we launch the debugger with an `attach` request type, VS Code will show a **Disconnect** button in the navigation bar. This button will disconnect the debugger from the app loaded in the browser. Using the request type (`launch`) will replace the disconnect button with a **Stop** button. The disconnect and stop buttons are shown here:

Figure 6.24 – Disconnect button

Figure 6.25 – Stop button

We have now explored the different navigation features provided by VS Code. Next, let's look at **variables**.

Inspecting data with the variables feature

While debugging code, you can view variables with their values in a nice structured view. The local variables list shows variables available in the class or method, based on the context. The following screenshot shows the variables available inside the `handleSuccess` method:

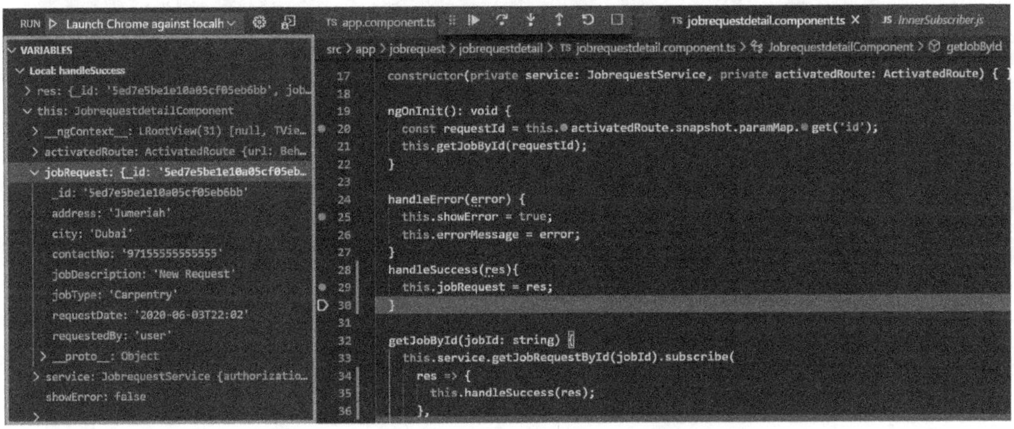

Figure 6.26 – The VARIABLES section

You can also set the value of a variable by either double-clicking it or right-clicking and selecting **Set Value** from the context menu, as shown here:

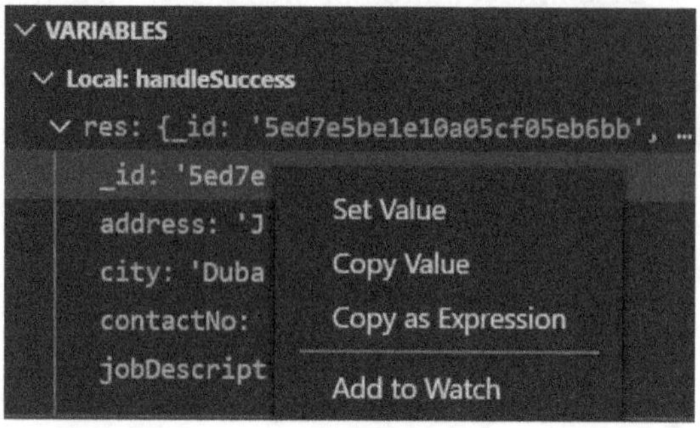

Figure 6.27 – Setting a value for a variable

Using the **Add to Watch** option, you can add a variable to the **WATCH** section.

Short and sweet, but the data inspection feature comes in handy while debugging. Next is the **call stack**.

Looking at the CALL STACK

The **CALL STACK** section shows the sequence of calls the application has made, with the top element showing the current program being debugged, as illustrated in the following screenshot:

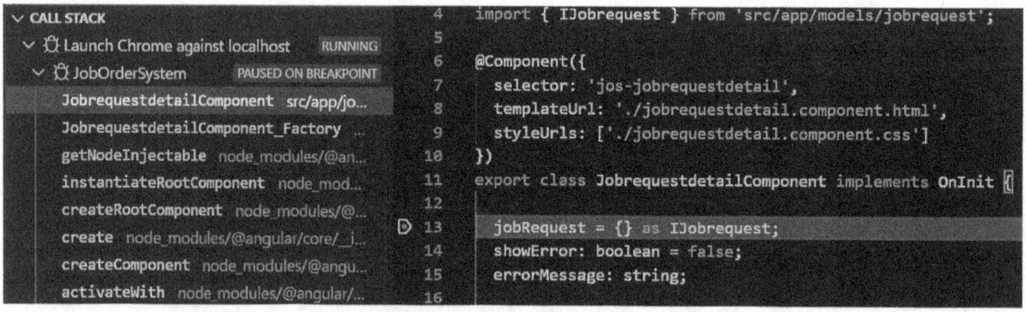

Figure 6.28 – CALL STACK

Figure 6.28 shows `JobrequestdetailComponent` at the top of the call stack. This component is currently being debugged in the right-hand window.

Exploring logpoints

Often, developers use console logs for debugging programs. Logpoints provide an option to easily log messages to the debug console. These are represented by a ◆ diamond sign in VS Code.

You can add a logpoint similarly to a breakpoint. Use the right mouse button to click next to a line number. The following screenshot shows the context menu with an **Add Logpoint...** option:

Figure 6.29 – Adding a logpoint

Logpoints can also be used to log values from a variable. Enclose the variable name in curly braces { } to print its value. VS Code provides IntelliSense for searching variables, as shown in the following screenshot:

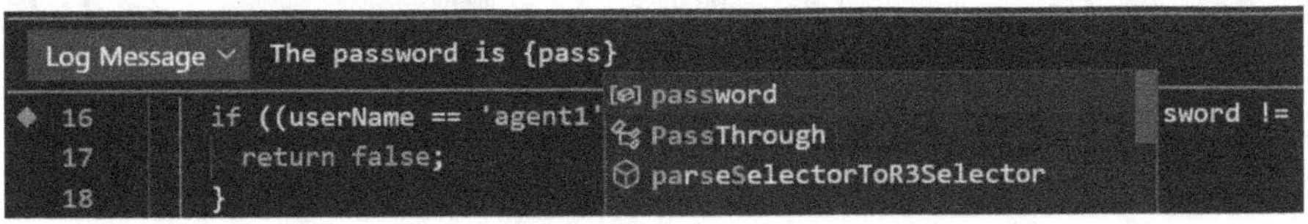

Figure 6.30 – Printing the password variable to the console

Once the shown logpoint is reached, the debugger will print the message along with the value of the password variable in the debug console, as shown here:

Figure 6.31 – Message logged with logpoint

As shown, the debugger logs the message to the debug console. Here, the name of the variable is replaced with its value—that is, `userpass`.

Debugging with split screen

VS Code allows you to split the screen and edit multiple files at the same time. These features also come in handy while debugging code where the program is internally calling methods or functions. Instead of switching between files, a split window makes it easy to debug on the same screen, as illustrated in the following screenshot:

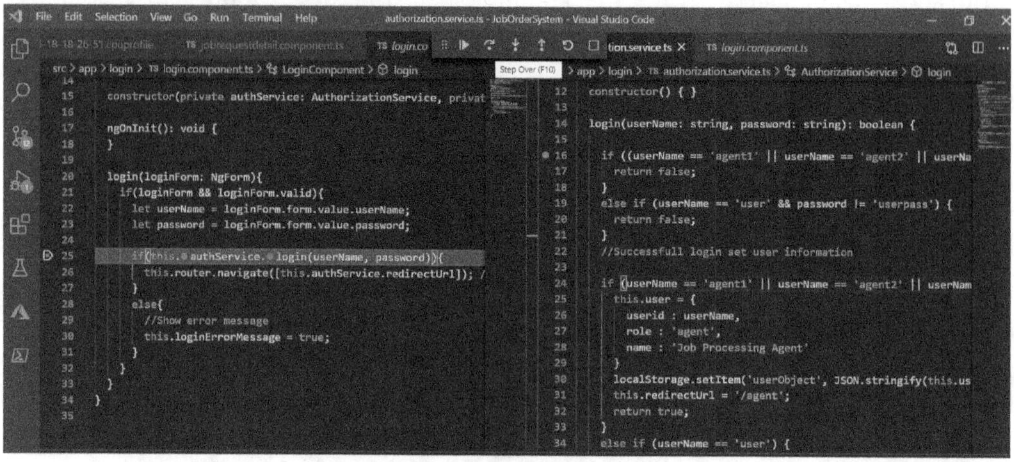

Figure 6.32 – Debugging with split screen

In the example shown in *Figure 6.32*, `LoginComponent` calls the `login` method of `AuthorizationService`. Opening both these files side by side will move the cursor between the two files. In *Figure 6.32*, the debugger stops at the `if` condition in the `LoginComponent`. To debug the `login` method of the `authorization service`, the cursor then moves to the file on the right in the `AuthorizationService` login method.

In this section, we explored several debugging features of VS Code using the Angular app. Next, let's move on to configure the debugger for the Node.js API project.

Debugging the Node.js API

While debugging the Angular app, we have focused on exploring the debugging extension for Chrome and also explored several different debugging features provided by VS Code. Apart from the Angular frontend, the job order system is spread over multiple microservices developed in different languages. Let's discuss how we can debug these multiplatform backends, starting off with Node.js APIs.

VS Code provides out-of-the-box support for Node.js. It's possible to debug code written in JavaScript and TypeScript without the need to install third-party extensions.

Creating the launch configuration

To set up the launch configuration for the Node.js project, we will follow the same approach and create a launch configuration, like this:

1. Open the Node.js project folder and press *Ctrl + Shift + P* to open the command palette. Type `Open launch.json`, as shown in the following screenshot, and press *Enter* to proceed:

Figure 6.33 – Create a launch configuration for the Node project

2. VS Code will prompt you to select an environment. Here, select **Node.js**, and VS Code will generate a `launch.json` file in the `.vscode` folder, as illustrated in the following code snippet:

```
{
        'type': 'node',
        'request': 'launch',
        'name': 'Launch Program',
        'skipFiles': [
            '<node_internals>/**'
        ],
        'program': '${workspaceFolder}\\server.js'
}
```

Let's discuss the launch configuration. The type for a Node project is `node`, and we are using `launch` as the request type. The name specifies the name to be displayed in the **Launch** dropdown in the debugger top bar. `skipFiles` specifies the files to be excluded from debugging. So, what does the `program` parameter do? Recall that we kick-start our Node server by running `npm start`. This command looks at the `package.json` file and runs the command specified in the `start` parameter, as shown in the following screenshot:

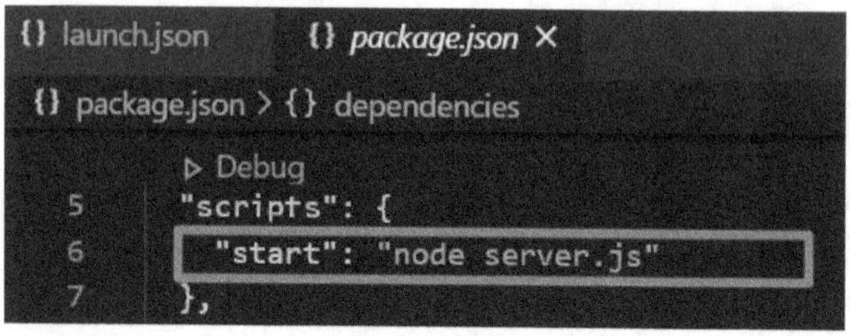

Figure 6.34 – Command linked with npm start

In the `program` parameter, we specify the same file. The launch configuration type is `node`, which means that when the debugger is launched, it will kick-start the node server using `server.js`.

The configuration is complete, so launch the debugger using *F5*. The debug console will show that the server is listening on port `3001`, and the bottom bar in VS Code will turn red. To test whether everything works fine, put a breakpoint in the root path, where the server responds with a welcome message, as shown in the following screenshot. After this, go to the browser and hit `http://localhost:3001`:

Figure 6.35 – Debugging the Node.js API

As shown in *Figure 6.35*, the breakpoint we put earlier is called, and control is transferred to the debugger.

Attaching the debugger

Another way to debug using VS Code is to attach the debugger to a running process. For Node.js, there are multiple ways of doing this—let's go through them in detail.

Auto Attach

To attach VS Code to a running Node server, you can use the **Auto Attach** option of VS Code. To enable this option, call the command palette with *Ctrl + Shift + P* and type `Toggle Auto Attach`. This is the Node debug **Auto Attach** extension. You will see the **Auto Attach** button enabled in the status bar, as follows: `Auto Attach: On`. Once this button is visible, you can also use it to turn the **Auto Attach** option on and off.

Next, we launch the `node` server in `inspect` mode, using the following command. It must be run in the VS Code integrated terminal:

```
node --inspect server.js
```

Alternatively, you can use the following command. This will make sure the debugger is attached before the program starts running:

```
node –inspect-brk server.js
```

The VS Code status bar will turn red, and the debugger navigation bar will appear at the top of the screen. Go to any browser and hit the **Uniform Resource Locator** (**URL**). Control will be transferred to VS Code for debugging.

Attaching to an existing process

Another option is to attach the debugger to an existing node process. These processes also include ones that are triggered from outside VS Code.

Let's try from outside VS Code. Open Command Prompt/a Terminal window and run the following command in the Node project directory:

```
node server.js
```

The server will spin up and start listening on port `3001`, as shown in the following screenshot:

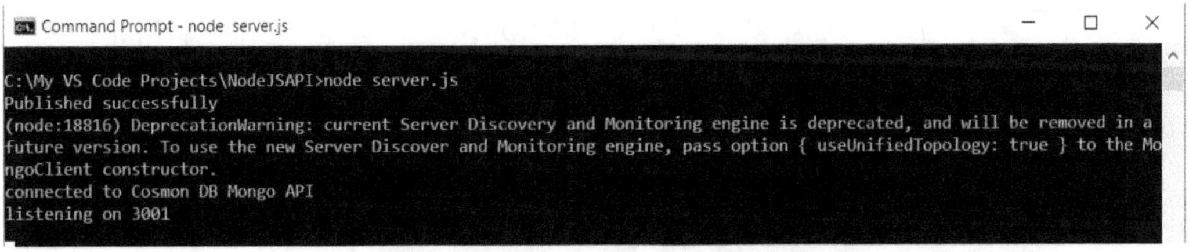

Figure 6.36 – Starting the Node server from a terminal window outside VS Code

To attach VS Code to this node process, call the **Command Palette** and look for **Attach to Node Process**, as illustrated in the following screenshot:

Figure 6.37 – Attach to Node Process from VS Code

VS Code will show a list of node processes running in your machine. Select the process to link the debugger, as illustrated in the following screenshot:

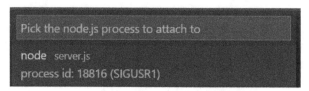

Figure 6.38 – Node processes running

On selecting the process, VS Code will attach the debugger to an already running Node process. The status bar will turn red, and, as shown in the following screenshot, the terminal window will show a `Debugger attached` message:

```
listening on 3001
Debugger listening on ws://127.0.0.1:9229/85a9f109-18ad-4adf-91d1-e9be1455bad6
For help, see: https://nodejs.org/en/docs/inspector
Debugger attached.
```

Figure 6.39 – Debugger attached to an existing process

Hitting the URL from any browser will call the debugger in VS Code.

From our overall application architecture, we have now covered the Angular application and also the Node.js API. Next, let's look at debugging our Java project with VS Code.

Debugging the Java API

Similar to the previous projects, we start by creating the launch configuration. Call the command palette with *Ctrl + Shift + P* and select **Debug: Open launch.json**. In the environment, select **Java**, as shown in the following screenshot:

Figure 6.40 – Create a launch configuration for Java

VS Code will create the launch configuration with two options. Apart from the `mainClass` parameter, the launch configuration is similar to our previous examples. In the first option, we tell the debugger to use the current active file, as illustrated in the following code snippet:

```
{
        'type': 'java',
        'name': 'Debug (Launch) - Current File',
        'request': 'launch',
        'mainClass': '${file}'
},
```

The second option explicitly defines the main class, which in our case is the `DemoApplication.java` file, as illustrated in the following code snippet:

```
{
        'type': 'java',
        'name': 'Debug (Launch)-DemoApplication<schedule>',
        'request': 'launch',
        'mainClass': 'com.jobsystem.schedule.
        DemoApplication',
        'projectName': 'schedule'
}
```

Select the launch configuration and run the debugger. VS Code will spin up the Java server and attach the debugger.

VS Code supports a variety of languages through its extension framework. Next on our list is Python.

Debugging the Python API

By now, we are done with the Angular app, Node.js, and the Java API. Next, let's look into setting up debugging for our Python project. Open the project folder in VS Code and go to the **debug** tab. Create the `launch.json` file. As shown in the following screenshot, select the environment as **Python**:

Figure 6.41 – Select a Python environment in the command palette

We have built our API using Flask, so select **Flask**, as shown in the following screenshot:

Figure 6.42 – Select Flask

VS Code will generate the following launch.json file in the .vscode folder:

```
{
            'name': 'Python: Launch',
            'type': 'python',
            'request': 'launch',
            'module': 'flask',
            'stopOnEntry': true,
            'env': {
                'FLASK_APP': 'app.py',
                'FLASK_ENV': 'development',
                'FLASK_DEBUG': '0'
            },
            'args': [
                'run',
                '--no-debugger',
                '--no-reload'
            ],
            'jinja': true
}
```

Let's walk through the parts of the launch configuration. This time, the type is python, and the module is flask. The FLASK_APP parameter specifies our app file, which holds the APIs.

You will notice a new `stopOnEntry` parameter, which is set to `true`. This will stop the execution and will transfer control to the debugger as soon as the first line of code is executed, as illustrated in the following screenshot:

Figure 6.43 – Debugger breaks at the first line of code

Figure 6.43 shows that, on launch, the debugger stops at the first line in our `app.py` file.

With Python, we have now come to the last component of our overall application. We have already discussed Angular, Node.js, Java, and Python. Let's look at .NET Core next.

Debugging .NET Core

The .NET Core service manages events triggered in Kafka. To debug the .NET Core application, we will first need to install the C# extension from the **MARKETPLACE**. Search for **C#**, as shown in the following screenshot:

Figure 6.44 – C# extension

Similar to previous projects, we will create a `launch.json` file from the **debug** tab. In the environment, select **.NET Core**, as shown in the following screenshot:

Figure 6.45 – Set the debug environment as .NET Core

VS Code will create two files, `launch.json` and `tasks.json`, as shown in the following screenshot:

Figure 6.46 – Launch configuration for the .NET Core API

In the launch configuration, the `type` is specified as `coreclr`; the request `type` is `launch`; `program` specifies the project's **Dynamic Link Library (DLL)** file; and `cwd` specifies the working directory. The code can be seen in the following snippet:

```
{
        'name': '.NET Core Launch (web)',
        'type': 'coreclr',
        'request': 'launch',
        'preLaunchTask': 'build',
        'program': '${workspaceFolder}/bin/Debug/netcoreapp3.1/
        NetCoreAPI.dll',
        'args': [],
        'cwd': '${workspaceFolder}',
        'stopAtEntry': false,
        'serverReadyAction': {
        'action': 'openExternally',
        'pattern': '\\bNow listening on:\\s+(https?://\\S+)'
        },
```

If your app has settings specific to a development or production environment, the `env` parameter can be used for specifying the environment variable, as follows:

```
'env': {
                'ASPNETCORE_ENVIRONMENT': 'Development'
        },
```

Our current application does not contain views. In the case of `Views`, the `sourceFileMap/Views` parameter points to the folder containing the views, as illustrated in the following code snippet:

```
'sourceFileMap': {
        '/Views': '${workspaceFolder}/Views'
```

```
        }
}
```

There is a new `preLaunchTask` parameter here, which tells VS Code to execute the `task` labeled as `build` in the `tasks.json` file before the debugger is launched. The following code snippet shows the `build` task configuration from the `tasks.json` file:

```json
{
        'label': 'build',
        'command': 'dotnet',
        'type': 'process',
        'args': [
            'build',
            '${workspaceFolder}/NetCoreAPI.csproj',
            '/property:GenerateFullPaths=true',
            '/consoleloggerparameters:NoSummary'
        ],
        'problemMatcher': '$msCompile'
},
```

The `build` tasks run the `dotnet` command, passing `build` and the `NetCoreAPI.csproj` project file as arguments. This is similar to running `dotnet build` in your project directory using a terminal window.

In the `launch.json` file, you will notice that VS Code has also generated the configuration for the `attach` request type. When the debugger is launched, the `processId` parameter requests VS Code to show a list of processes in the command palette, as illustrated in the following code snippet:

```json
{
        'name': '.NET Core Attach',
        'type': 'coreclr',
        'request': 'attach',
        'processId': '${command:pickProcess}'
}
```

As shown in the following screenshot, you can select your specific **dotnet** program to attach the debugger:

Figure 6.47 – List of processes to attach the debugger

Finally, with the .NET Core project, we have completed setting up and exploring VS Code debugging features and options for different languages used in our job order system application. You must be amazed by how such a sleek and lightweight editor has the capability to extend and cover such a diverse portfolio of platforms and languages.

Summary

In this chapter, we started by discussing the debugger layout and its various sections. Keeping a focus on what was developed in previous chapters, we looked at each project and discussed how VS Code can cater to its debugging requirements. During this chapter, we learned how to configure VS Code for debugging Angular, Node.js, Java, Python, and .NET Core projects. We also learned about several debugging features, such as breakpoints, variables, the call stack, and logpoints, and we discussed the navigation features as well.

In the next chapter, we will be containerizing the services and frontend app created for the job order system application. We will provision the Azure Kubernetes cluster and deploy the containerized services on Azure.

7
Deploying Applications on Azure

Today, there are various hosting options available that can be used to deploy applications. The application can be hosted on a physical machine, a virtual machine in the cloud, on on-premises servers, or through leveraging managed services in the cloud. The traditional approach of setting up a website is to spin up a virtual machine or a physical machine, and then configure the web server and deploy the application on that web server. On the other hand, we can leverage managed services in Azure such as Azure App Service, Azure Kubernetes Service, and Azure Container Instances. With managed services, you need to focus more on the application rather than the infrastructure. The cloud platform offers high scalability and availability options and a better **service level agreement (SLA)**, which is comparatively harder to achieve with on-premise solutions.

Azure App Service is a fully managed web hosting platform used to build web applications, web APIs based on RESTful standards, or connected mobile backends. Since it is a fully managed PaaS service, it encapsulates the complexities of spinning up the VM, setting up the web server, and configuring the application on it. Being a developer, you just need to focus on the application and trigger a deployment, where the deployment-related activities are done by App Service itself. Azure App Service supports multiple languages and frameworks and provides first-class support for ASP.NET, ASP.NET Core, Java, and many other platforms. Applications deployed as an App Service can easily scale out and provide high availability.

We can also deploy containers in Azure App Service, though there are specific platforms available in Azure that provide complete container orchestrating features. In Azure, we can use managed Kubernetes services known as **Azure Kubernetes Services** to deploy applications running inside containers. It is a very powerful platform that's used to orchestrate containers running on Azure. It provides lots of managed features such as scaling out clusters, scaling out pods, and monitoring the performance of your applications running on containers.

In this chapter, we will be discussing the following topics:

- Setting up the development environment to work with Docker and Kubernetes
- Understanding the importance of containers and using Docker as a container technology
- Building Docker images for microservices developed in Node.js, the Java Spring Boot API, Python, and .NET Core, and developing the frontend application using the Angular framework
- Provisioning an Azure Container Registry on Azure using VS Code
- Deploying Docker images to Azure Kubernetes Service

Technical requirements

To containerize applications, we need to set up a development environment and install Docker and Kubernetes tools. We'll learn how to do this in the following sections.

Installing Docker

Docker is an open source platform for developing, shipping, and running applications. Docker runtime is needed when working with Docker containers. To install Docker for the respective operating system, download it from the following link:

```
https://docs.docker.com/get-docker/
```

Once the Docker runtime has been installed, you can verify installation by running the following command:

```
docker --version
```

This command lists the version number of the Docker runtime.

Installing the Kubernetes CLI

The Kubernetes CLI is a command-line tool that allows you to run commands against Kubernetes clusters. You can use kubectl commands to manage clusters, create deployments, deploy pods and services, and so on.

You can install kubectl on Windows by referring to the following link:

```
https://kubernetes.io/docs/tasks/tools/
install-kubectl/#install-kubectl-on-windows
```

Once kubectl has been installed, you can verify the installation by running the following command:

```
kubectl --help
```

The preceding command lists the command used to manage a Kubernetes cluster.

Now that the necessary tools have been installed, we can learn about the importance of containers and why they serve better than VMs.

Why containers?

A container is a standard application unit that bundles code and related dependencies and runs autonomously in a sandbox environment. With containers, you get the complete isolation as a virtual machine; however, it is more lightweight and abstraction is done at the **operating system (OS)** level. Containers give developers the ability to run applications in different containers within the same machine with no dependencies in terms of installing software components or runtimes. Each container can run different applications based on different platforms and have different dependencies. Containers provide OS-level virtualization, whereas VMs are based on hardware-level virtualization.

The following diagram shows a comparison of VMs and containers:

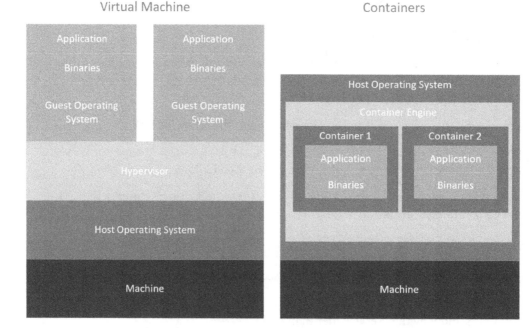

Figure 7.1 – Comparison of virtual machines and containers

In the preceding diagram, we can see that VM virtualization is done at the **Hypervisor** level, whereas for containers, virtualization is done at the **OS** level.

Some of the benefits containers provide over a VM or physical machine include zero downtime, easy to upgrade, faster start up time, the need for less memory space, and the provision of process-level isolation, which amounts to an ideal solution for microservices.

There are various container technologies available, such as Docker, Rancher, Apache Mesos, and many others. However, we will use Docker as the underlying technology for containerizing our services and as the frontend application for our **Job Order System** (**JOS**). We'll have a brief look at Docker in the next section.

Docker containers

Docker is a technology used to containerize applications. Like Docker, there are many alternatives, such as Apache Mesos, Rancher, Virtual Box, and others. However, Docker is one of the most widely adopted technologies for building and running containers for applications. Regardless of the technology, you can create Docker images for any application that can run on any platform.

The following are some of the core characteristics of a Docker container:

- Docker is a standard in the industry for containerization and can build applications on any technology or platform.

- Each docker container shares the kernel of the host OS and does not require that you have a separate OS.

- Provides the most robust isolation capabilities available compared to other container technologies.

Docker provides containers for both Windows and Linux distros. On the host OS where the Docker runtime is installed, you can switch between Linux and Windows containers and build Docker images for your applications based on the target OS you wanted to host them. Let's explore how to build the Docker images for a JOS.

Building Docker images

In a **JOS**, we have multiple microservices we can use to create job requests, schedule job requests, and assign agents so that they send notifications, listen for Kafka events, and listen to a frontend application. In this section, we will be containerizing them using Docker.

VS Code provides a Docker extension that makes it easy for developers to build Docker files for any platform. The Docker extension can be installed from the **EXTENSIONS** tab, as shown in the following screenshot:

Figure 7.2 – Installing the Docker extension

Once the Docker extension has been installed, you can open the command palette and select the Docker command, which shows various options you can use to choose your desired language or platform. It will take you through the series of steps to create the respective Dockerfile for that platform:

Figure 7.3 – Adding a Dockerfile

A Dockerfile is the main file used to create docker images. It contains commands to create an image. It also contains information about the base image, dependencies, the application package to be built or referenced, port information where the application will run, and the entry point or command to spin up the application.

Now, let's build some Docker images for all the services and frontend applications of JOS, which are as follows:

- Job Request Service
- Schedule Job Service
- Notifications Service
- Hosted Service
- JOS Web Application

We'll start with the Job Request Service.

Building a Docker image for the Job Request Service

The Job Request Service is built on Node.js.

To create a Dockerfile for a Node.js Express application, open the command palette and select the `docker: add docker` command.

Once the command has been selected, a list of languages or platforms will appear. From here, do the following:

- Select **Node.js** and proceed with the wizard to create a docker file:

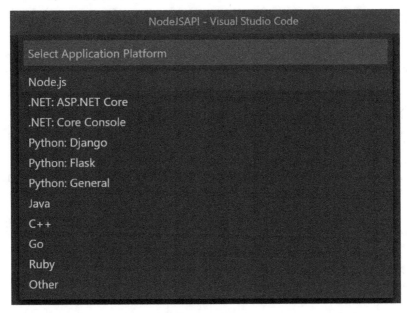

Figure 7.4 – Selecting Node.js as the application platform

- Once this `Dockerfile` has been added to the Job Request service workspace, you can verify and even customize it based on your needs. For example, the docker file we created out of the box may not have the proper dependencies. Here is our `Dockerfile` for the Job Request service:

```
FROM node:10.13-alpines
RUN mkdir -p /usr/src/app
WORKDIR /usr/src/app
COPY . /usr/src/app
RUN apk --no-cache --virtual build-dependencies add \
    python \
    bash \
```

```
      make \
      g++ \
      && npm install \
      && apk del build-dependencies
EXPOSE 3001
CMD npm start
```

The structure of the docker file starts with the base image. We've used node version 10 of an Alpine image as the base image to run the Job Request Service. Then, we specified /usr/src/app as the command that will be used to create the directory. The -p switch is used to create a directory with a missing parent directory.

- Next, we set the working directory to the directory we just created inside the container and copy all the files from the root directory from the local machine to the newly created folder insider the container.

- Since the Node.js API has several dependencies, we need to install the dependencies using the RUN apk instruction and mention all the dependencies by using the add switch.

- Finally, we expose the port as 3001 and execute the run command using the CMD instruction.

> **Note**
>
> The docker extension is intelligent enough to pull the correct base image, as per the language version or framework you are using, and create the *Dockerfile* accordingly.

Once the docker file has been created, the container image can be built using the following command:

```
docker build -t {image_name:version_no} {folder_path_to_
dockerfile}
```

For the Job Request API, we can name the image jobrequestservice and the initial version 1.0. The -t option is used to tag the image with the given name in the command. image_name is the name of the application, while version_no is the version of the image you are going to build. Following version_no, you can put a period, ., followed by the current path or specify a complete path where the Dockerfile resides.

Here is the actual docker command we used to build the image for the Job Request Service. While running this, make sure you are in the folder path where the Dockerfile resides:

```
docker build -t jobrequestservice:1.0 .
```

After running the preceding command, it searches for the Dockerfile on the current path from where the command is executing and starts executing the steps in sequence, as defined in the Dockerfile.

The following screenshot shows the steps that are executed when the docker `build` command is executed:

```
PS C:\Books\VSCode\NodeJSAPI> docker build -t jobrequestservice:1.0 .
Sending build context to Docker daemon  81.41kB
Step 1/8 : FROM node:10.13-alpine
10.13-alpine: Pulling from library/node
4fe2ade4980c: Pull complete
c245f6a8ecc5: Pull complete
82bdc9503d50: Pull complete
Digest: sha256:22c8219b21f86dfd7398ce1f62c48a022fecdcf0ad7bf3b0681131bd04a023a2
```

Figure 7.5 – Building a Docker image

Once the image has been built, we can verify the creation of the image by running the following command:

```
docker images
```

The preceding command shows the list of docker images that exist in the system. Please note that the container still hasn't been created yet. To create the container and access the service, we need to run the following docker `run` command:

```
docker run -p 3001:3001 jobrequestservice:1.0
```

Once this command has been executed, a container will be created and we can access the docker image from the local IP address or by using localhost, as follows:

Figure 7.6 – Access Job Request API on port 3001

You can access the other methods, such as /jobs, with HTTP GET and POST requests, which will allow you to perform read and write operations.

Building a Docker image for the Schedule Job Service

The Schedule Job Service is built on top of the Java Spring Boot API framework.

First, we need to build a package (a .jar file) for the Schedule Job Service. By doing this, we can refer to this package while configuring the Dockerfile. To build the package, we can run the following command:

```
mvn package
```

While running this command, make sure you are at the root of the schedule folder.

To create a Dockerfile for a Java application, we will repeat the same steps of opening the command palette , selecting the docker: add docker command and selecting Java from the Application Platform dialog. Next, it will ask for the application port and create a Dockerfile. Refer to the following screenshot:

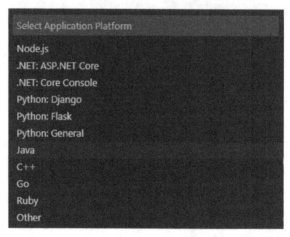

Figure 7.7 – Selecting Java as the application platform

Once the Dockerfile has been added to the Schedule Job API workspace, you can verify and even customize it based on your needs.

Here is the complete `Dockerfile` for the Schedule Job Service:

```
FROM openjdk:8-jdk-alpine
VOLUME /tmp
ARG JAVA_OPTS
ENV JAVA_OPTS=$JAVA_OPTS
ADD target/schedule-0.0.1-SNAPSHOT.jar schedule.jar
EXPOSE 9003
ENTRYPOINT exec java $JAVA_OPTS -jar schedule.jar
# For Spring-Boot project, use the entrypoint below to reduce
Tomcat startup time.
#ENTRYPOINT exec java $JAVA_OPTS -Djava.security.egd=file:/
dev/./urandom -jar schedule.jar
```

From the preceding code block, we can see that the Dockerfile starts from the base image, which is the Alpine version of the Java SDK. The next statement specifies the working directory where the files will be copied using the VOLUME instruction. The ARG command is used to specify any arguments that will be passed when building the docker file. Then, we added the `schedule.jar` file, which contains the actual package we built previously. Finally, the port that will be used is 9003 and an ENTRYPOINT instruction is used to run the Java application.

For the Job Schedule Service, we can name the image jobrequestservice and the initial version 1.0. The -t option is used to tag the image with the given name in the command.

Here is the actual docker command we used to build the image for the Schedule Job Service:

```
docker build -t schedulejobservice:1.0 .
```

Running the preceding command will create a docker image named schedulejobservice. To run this image, we can use the following command:

```
docker run -p 9003:9003 schedulejobservice:1.0
```

Once the container has been created, we can access it using the local IP address or localhost from the host machine where docker runtime is running.

Building a Docker image for the Notifications Service

The Notifications Service is built on Python.

To create a `Dockerfile` for a Python application, we can repeat the same steps of opening the command palette and selecting the `docker: add docker` command that we completed previously. This time, however, in the application platform selector dialog, select the `Python: Flask` option and proceed. This creates a docker file that looks as follows:

```
FROM python:3.8-slim-buster
EXPOSE 5000
# Keeps Python from generating .pyc files in the container
ENV PYTHONDONTWRITEBYTECODE 1
# Turns off buffering for easier container logging
ENV PYTHONUNBUFFERED 1
# Install pip requirements
ADD requirements.txt .
RUN pip install sendgrid
RUN python -m pip install -r requirements.txt
WORKDIR /app
ADD . /app
RUN useradd appuser && chown -R appuser /app
USER appuser
# During debugging, this entry point will be overridden.
CMD ['gunicorn', '--bind', '0.0.0.0:5000', 'app:app']
```

From the preceding code, we can see that the structure of the docker file starts with the base image. Let's take a look at this in more detail:

- The `EXPOSE` instruction is used to mention the port number where the application will be configured to listen at.

- `PYTHONDONTWRITEBYTECODE` is used to enable or disable Python from writing .pyc files on the imports of source modules. If it is set to a non-empty string, Python won't try to write files.

- If `PYTHONUNBUFFERED` is specified in the non-empty string, it forces the `stdout` and `stderr` streams to be unbuffered.

- `ADD requirements.txt` is used to read the `requirements.txt` files for all the modules and install them. However, for `sendgrid`, we explicitly mentioned `pip install` for the `sendgrid` module.

- The working directory we created inside the container has been configured using the `WORKDIR` instruction, where `/app` is the actual folder. Then, we copied the files from root to the `/app` folder using the `ADD` keyword.

- The `useradd` command, along with the `RUN` instruction, is used to switch a non-root user and change the ownership of the `/app` folder.

- Finally, the container can be run using the `CMD` instruction.

For the Notifications Service, we can name the `notificationsservice` image and build it using the following command:

```
docker build -t notificationsservice:1.0 .
```

The preceding command traverses the Dockerfile and executes the steps in sequence. Finally, the image will be created and can be run using the docker `run` command, as follows:

```
docker run -p 5000:5000 notificationsservice:1.0
```

Once the container has been created, the Notifications Service can be accessed from the host machine on port `5000`.

Building a Docker image for the Hosted Service

The Hosted Service is built on top of .NET Core. This is a background service that listens to Kafka events and call services for integration. To create a `Dockerfile` for a .NET Core application, we need to repeat the same steps of opening the command palette and selecting the `docker: add docker` command that we followed previously. This time, in the `Application Platform` selector dialog, select the `.NET: Core Console` option and proceed with the steps shown in the wizard to create a `Dockerfile`.

Here is the Dockerfile for the .NET Core Hosted Service:

```
FROM mcr.microsoft.com/dotnet/core/aspnet:3.1 AS base
WORKDIR /app
FROM mcr.microsoft.com/dotnet/core/sdk:3.1 AS build
WORKDIR /src
COPY ['NetCoreAPI.csproj', './']
RUN dotnet restore './NetCoreAPI.csproj'
```

```
COPY . .
WORKDIR '/src/.'
RUN dotnet build 'NetCoreAPI.csproj' -c Release -o /app/build
FROM build AS publish
RUN dotnet publish 'NetCoreAPI.csproj' -c Release -o /app/
publish
FROM base AS final
WORKDIR /app
COPY --from=publish /app/publish .
ENTRYPOINT ['dotnet', 'NetCoreAPI.dll']
```

From the preceding code, we can see that the structure of the docker file starts from the base image of ASP.NET Core 3.1. Let's take a look at this in more detail:

- Here, we set the working directory in the container using the WORKDIR instruction, specified it as /app, and used base as an alias.

- To build a Hosted Service inside a container, we need to refer to the SDK image of .NET Core and give it a build alias. After doing this, we set a different directory, / src, to build the application package inside it.

- We then copied the project file using the COPY instruction to the root folder of /src and ran the dotnet restore command to read the project file and download all the dependencies. Once the packages had been restored, we copied all the other files and placed them inside the /src folder using the COPY . . instruction.

- Next, we built the application package using the dotnet publish command and set the output directory as /app/publish. In this statement, the directory will be created inside the src folder and named app. Inside app, there will be a publish folder that contains the application's published files.

- Then, we picked up the base image and set the working directory to the /app folder, which is where the publish folder resides, and copied all the files from /app/publish to the root of the /app folder.

- Lastly, we ran the application using the ENTRYPOINT instruction.

To build the image for the Hosted Service, run the following command:

```
docker build -t hostedservice:1.0 .
```

The preceding command reads the Dockerfile and executes the steps in sequence. On successful execution, the image will be created, and we can run the image using `docker run`, as follows:

```
docker run -p 6000:6000 hostedservice:1.0
```

In this section, we built docker images for all the services in our JOS application. Our backend services are now running inside containers. In the next section, we will configure a Docker image for the JOS Web App and build a docker image.

Building a Docker image for the JOS Web App

We know that the frontend for JOS is built on Angular. Since Angular runs on Node.js, we can use the similar approach of creating a Dockerfile using the command palette, just like we did when creating a `Dockerfile` for the Job Request service.

Here is the Dockerfile for the JOS frontend application:

```
# base image
FROM node:12.2.0
# set working directory
WORKDIR /app
# add `/app/node_modules/.bin` to $PATH
ENV PATH /app/node_modules/.bin:$PATH
# install and cache app dependencies
COPY package.json /app/package.json
RUN npm install
RUN npm install -g @angular/cli@9.1.4
# add app
COPY . /app
# start app
CMD ng serve --host 0.0.0.0
```

From the preceding code, we can see that, for the JOS frontend app, we used node as the base image and then set the working directory to the /app folder. After this, we did the following:

- We then set app/node_modules as PATH in the environment variables of the container and copied the package.json file, which contains the dependencies that need to be installed.

- The RUN npm install command reads the package.json file and installs all the dependencies required. It then adds them under the node_modules folder, inside the app folder in our container.

- Finally, we installed Angular globally on the container machine so that the ng serve command can be executed and copied all the files from root to the /app folder.

To build the image for the JOS web app, run the following command:

```
docker build -t josapp:1.0 .
```

The preceding command reads the Dockerfile and executes the steps in sequence. On successful execution, the image will be created, and we can run the image using the docker run command, as follows:

```
docker run -p 4200:4200 josapp:1.0
```

In this section, we created and built docker images for all the services in our JOS application. Our backend services are now running inside containers.

So far, we have built docker images for all our services. The following table shows a summary of the images we've created, along with their ports:

Service Name	Docker Image Name	Port
Job Request Service	jobrequestservice	3001
Schedule Job Service	schedulejobservice	9003
Notifications Service	notificationsservice	5000
Hosted Service	hostedservice	6000

In the next section, we will configure a docker image for the JOS Web App and build it.

Setting up a private container registry in Azure

Docker images can be shared and deployed on Kubernetes or other orchestration engines through public or private registries. The docker public registry, known as *Docker hub*, can be used to store docker images. Moreover, Docker hub also allows you to provision private registries. However, a separate subscription needs to be purchased to use private registries in Docker hub.

In this section, since our solution is utilizing Microsoft Azure services, we will provision a new container registry on Azure known as the **Azure Container Registry** (**ACR**). The ACR provides a managed docker registry based on open source Docker registry 2.0, where you can store and manage docker images.

To provision the ACR from VS Code, we need to have the Docker extension installed. Once this has been installed, you can open the command palette and search for **Azure Container Registry**, as follows:

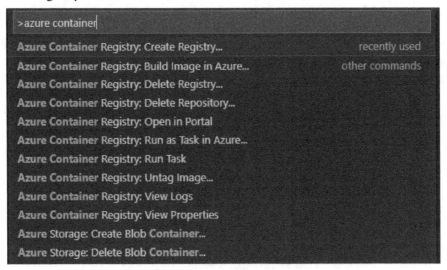

Figure 7.8 – Creating an Azure Container Registry instance

The wizard will ask you to specify a few details. Please enter the following values:

- Registry name: `josappregistry`
- SKU: `Basic`
- Resource Group: `VSCodeBookRG`
- Location: `West Europe`

You can also alter these values based on your usage. Proceeding through these steps will create the container registry in Azure.

To push images from your local machine to the ACR, we need to log in to the ACR and run the following two commands for all the images that need to be uploaded to the ACR:

- `Docker tag`: This is used to tag an image with the fully qualified name of the ACR.

- `Docker push`: This is used to push the tagged image to the ACR.

To log in to the ACR, execute the following command:

```
az login acr --name josappregistry
```

The `az` command comes with the Azure CLI, while the `--name` switch is used to mention the container registry's name.

Once the user is logged in, we can tag and push docker images to the ACR for all the services and frontend apps we created in this section:

1. To tag in VS Code, open the command palette, search for `Tag`, and select **Docker Images: Tag**, as shown in the following screenshot:

Figure 7.9 – Tagging an image

This will show a list of images that currently exist in the system. You will then be asked to choose one. Choose the service image:

Figure 7.10 – Selecting jobrequestservice

2. Then, choose the specific version you want to tag:

Figure 7.11 – Selecting the image version

3. Finally, choose the tagged name, which should be fully qualified, followed by the image name and version:

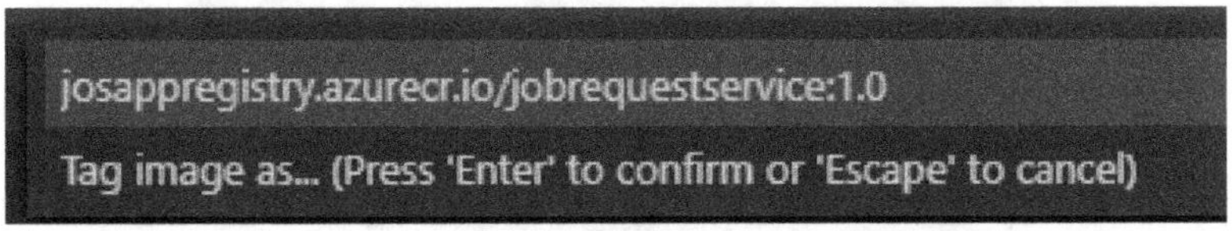

josappregistry.azurecr.io/jobrequestservice:1.0

Tag image as... (Press 'Enter' to confirm or 'Escape' to cancel)

Figure 7.12 – Access Job Request API on port 3001

Alternatively, images can be tagged using the Docker CLI. Here is the syntax to tag images:

```
docker tag {local_image_name:version_no} {fully_qualified_
registry_name}/{image_name}:{version_no}
```

To tag a Job Request Service, use the following syntax:

```
docker tag jobrequestservice:1.0 josappregistry.azurecr.io/
jobrequestservice:1.0
```

To tag a Job Schedule Service, use the following syntax:

```
docker tag schedulejobservice:1.0 josappregistry.azurecr.io/
schedulejobservice:1.0
```

To tag a Notifications Service, use the following syntax:

```
docker tag notificationsservice:1.0 josappregistry.azurecr.io/
notificationsservice:1.0
```

To tag a Hosted Service, use the following syntax:

```
docker tag hostedservice:1.0 josappregistry.azurecr.io/
hostedservice:1.0
```

To tag a JOS Web App, use the following syntax:

```
docker tag joswebapp:1.0 josappregistry.azurecr.io/
joswebapp:1.0
```

Once the images have been tagged, we can push them using VS Code by going to the command palette and choosing the **Docker Images: Push...** option, as shown in the following screenshot:

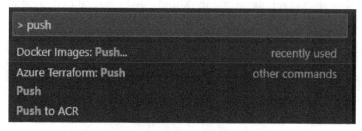

Figure 7.13 – Pushing images to a Container Registry

Then, we need to select the image we tagged previously:

Figure 7.14 – The name of the tagged image to push to the ACR

Alternatively, the images can also be pushed using the docker push command, as follows:

To push a Job Request Service, use the following command:

```
docker push josappregistry.azurecr.io/jobrequestservice:1.0
```

To push a Schedule Service, use the following command:

```
docker push josappregistry.azurecr.io/
schedulejobservice:1.0
```

To push a Notifications Service, use the following command:

```
docker push josappregistry.azurecr.io/notificationsservice:1.0
```

To push a Hosted Service, use the following command:

```
docker push josappregistry.azurecr.io/hostedservice:1.0
```

To push a JOS Web App, use the following command:

```
docker push josappregistry.azurecr.io/joswebapp:1.0
```

Once the images have been pushed, you can verify this from the portal by navigating to the ACR resource and selecting the `Repositories` tab.

Since the images have been pushed to the private registry, we will now explore Kubernetes and deploy them to Azure Kubernetes Service.

Deploying images to Azure Kubernetes Service

So far, you have containerized all the JOS services and are running docker containers locally. For production scenarios, we need to host them somewhere on the server. Let's say we need to scale out the Job Request Service, so we need to provision another VM with the Docker runtime and build and run that image there. After doing that, we need to add a load balancer that exposes an endpoint to take incoming requests and route that traffic back to the containers, depending on their availability, and so on.

There are various technologies available on the market that you can use to orchestrate docker containers for scalability. Of these, technologies such as Docker Swarm, Kubernetes, and Apache Mesos are the most popular ones. We will use Kubernetes and utilize a managed service of Kubernetes on Azure known as **Azure Kubernetes Service (AKS)** to accommodate this need.

Kubernetes is an open source, portable, and extensible platform for managing containerized applications. It was originally developed by Google and is now maintained as part of the **Cloud Native Computing Foundation (CNCF)**.

> **Note**
> CNCF is a project from the Linux Foundation that brings together the world's top developers, vendors, and end users to run the largest open source developer conferences in the world.

In the following sections, we'll learn about some key Kubernetes terminologies, as well as cluster representation and AKS.

Kubernetes terminology

It is important to understand what terminologies are used in the Kubernetes language. Some of the key terms that are used in Kubernetes are as follows:

- **Node**: A node is a virtual machine or physical machine where the containers are running. Each node in Kubernetes is part of the Kubernetes cluster.

- **Deployment**: Deployment specifies the replicas for the pods that need to be created. Deployments provide declarative changes for pods by describing a desired state.

- **Pod**: A pod is a building block where the containers run. Each pod has a private IP assigned to it and can have one or more containers running inside it.

- **Service**: The service is an access point for a pod. A service exposes a public endpoint and routes the ingress traffic to the underlying pod it is associated with.

You can learn more about other terminologies that are used in Kubernetes by going to the following link:

`https://kubernetes.io/docs/reference/glossary/?fundamental=true`

Now, let's learn more about Kubernetes and understand its high-level architecture.

Kubernetes cluster representation

The Kubernetes architecture is a collection of nodes, where a node is a virtual or physical machine that's part of the Kubernetes cluster. All the nodes are known as **worker nodes** and are managed by a master node.

The following diagram depicts the high-level architecture of a Kubernetes cluster:

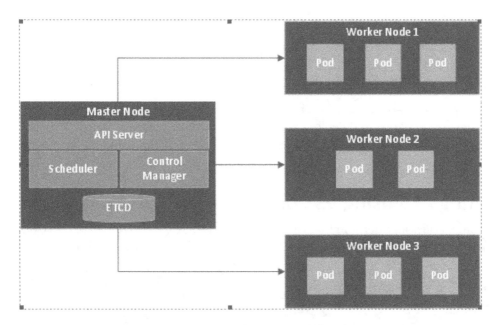

Figure 7.15 – A Kubernetes cluster

The **master node** contains the API server component that is used to receive all the commands that are executed through the kubectl tool. kubectl is the command-line tool that's used to run commands and manage Kubernetes. Container deployments, scaling the cluster, reading logs, and more can be done using the kubectl tool.

Whenever the user executes this command, the API server receives the request, validates it, and create objects in the **ETCD** database. For example, if we need to deploy the Job Request Service with two replicas in a Kubernetes cluster, we will create a YAML document for deployment and run the kubectl command to deploy it. The API server will get the request and create a record in **ETCD** that lets Scheduler create the deployment accordingly.

The **ETCD** is the distributed key-value data store used for service discovery, configuration management, and utilizing the state of Kubernetes objects, such as pods. In short, all the cluster's information is stored in the **ETCD**.

The **control manager** is a component that's used to watch the states of objects and make the necessary changes to move the current state to the desired state. For example, let's say we created a deployment that will run two pods for the Job Request Service. For some reason, one pod failed or dropped off. Here, the control manager immediately spins up a new pod on the available node by reading and comparing the current state with the desired state.

The scheduler is used to watch the pods that are not bound to any node and bind them. Once the pod has been assigned to a node, the pod and containers are created.

Azure Kubernetes Service (AKS)

AKS is a highly available, secure, and fully managed service on Azure that provides an easy setup and configures and deploys your containerized workloads to Azure. You can provision AKS in minutes from Azure Portal by using PowerShell, ARM templates, and a few other options.

Creating AKS using VS Code

Microsoft provides an extension that you can use to create AKS on Azure using VS Code. The extension can be installed from the **EXTENSIONS** tab in VS Code, as follows:

Figure 7.16 – Adding the Kubernetes extension

Once the extension has been installed, we can create a cluster, apply changes to the cluster, and use some other options. Let's see how this is done:

1. To create AKS, we need to access the command palette and search for `kubernetes`, as shown here:

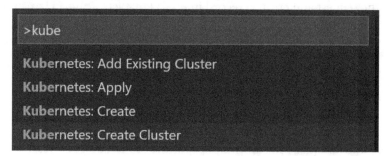

Figure 7.17 – Creating a Kubernetes cluster

2. Select the **Create Cluster** option and then choose **Azure Kubernetes Service** as the cluster type:

Figure 7.18 – Selecting Azure Kubernetes Service as the cluster type

3. Click **Next**, and then choose the desired subscription:

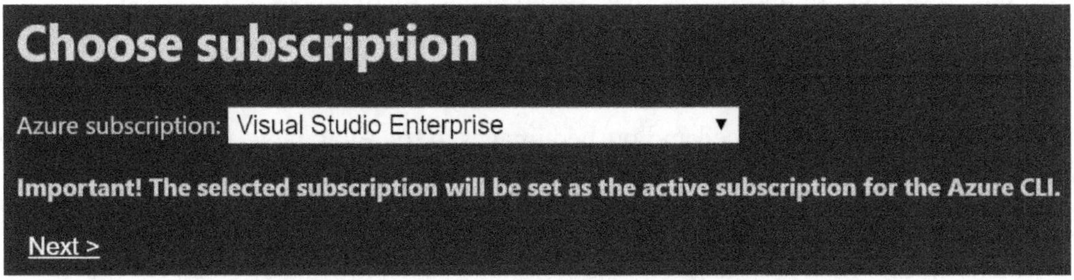

Choose subscription

Azure subscription: Visual Studio Enterprise ▼

Important! The selected subscription will be set as the active subscription for the Azure CLI.

Next >

Figure 7.19 – Selecting the desired Azure subscription

4. Specify **Cluster name**, **Resource group name**, and **Location**; this is where you need to provision AKS:

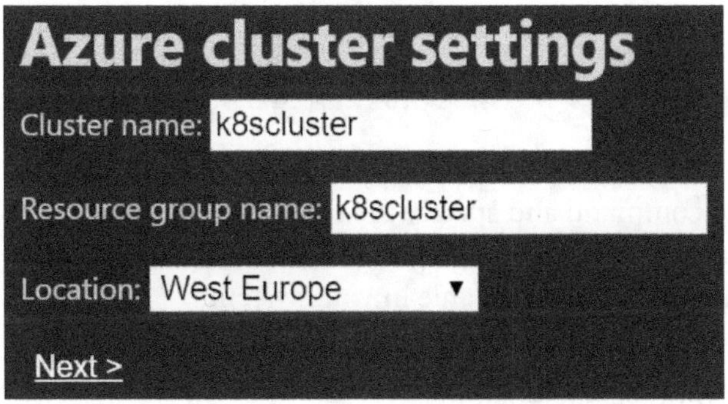

Azure cluster settings

Cluster name: k8scluster

Resource group name: k8scluster

Location: West Europe ▼

Next >

Figure 7.20 – Specifying Cluster name, Resource group name, and Location

5. Specify **Agent count**, which denotes the nodes and the corresponding number of virtual machines that you need to consider for the cluster. **Agent VM size** denotes the sizing plan in terms of CPU and memory:

Figure 7.21 – Selecting Agent Count and Agent VM size

6. Finally, click the **Create cluster** option, which creates the cluster on Azure.

In this section, we set up AKS on Azure using VS Code. In the next section, we will tag images and push them to the ACR so that we can deploy them on AKS.

Deploying tagged images from the ACR to AKS

In the previous section, we tagged and pushed images to the ACR. Now, we can use that image and deploy to AKS.

To create the pods for each microservice and deploy them on AKS, we need to create the deployment file and then use the `kubectl` command to deploy that on AKS. On the other hand, since a pod does not have a public IP address, we need to create a deployment of a service and create a service to bind the public IP. All the traffic will come to the service over a public endpoint, where it will then be routed to the respective pods.

To deploy images to the ACR, we need to get the necessary credentials for AKS. Run the following command to get the access credentials for our managed AKS instance:

```
az aks get-credentials --name k8cluster -resource-group
k8cluster
```

Modify the preceding command and specify the correct name for your AKS cluster, as well as the resource group. This command gets the necessary credentials from AKS and merges them inside the `kube.config` file in your system.

Now, create a deployment file and name it `JobRequest_K8_Deployment.yaml`. Here is the configuration for the deployment file:

```
apiVersion: apps/v1
kind: Deployment
metadata:
  name: jobrequestservice
```

```
spec:
  replicas: 2
  selector:
    matchLabels:
      app: jos
      component: jobrequestservice
  template:
    metadata:
      labels:
        app: jos
        component: jobrequestservice
    spec:
      containers:

      -
          image: 'josappregistry.azurecr.io/
jobrequestservice:1.0'
          name: jobrequestservice
          ports:

          -
              containerPort: 3001
      imagePullSecrets:
      - name: josappsecret
```

Starting from the top of the code, apiVersion is the version of the deployment kind, while Kind is the type of deployment we are configuring. Each kind of deployment contains two sections; namely, metadata and spec. metadata contains the name or labels we will use to reference other kinds of deployments. spec represents the actual attributes for a specific kind of deployment. As you may have noticed, spec contains replicas. The replicas keyword is used to specify how many pods we want to provision when this deployment is run.

The template keyword represents a Pod kind configuration. In the deployment kind, the configuration that's added after a template keyword is related to the Pod kind configuration. The Pod configuration contains the same metadata and spec sections that we have in the deployment kind. We can also create a separate Pod kind template and deploy certain pods. However, deployment ensures that all the pods have the correct configuration and that the right number of replicas (instances) are running.

For example, here is the sample `Pod` template that's used to create pods:

```
apiVersion: v1
kind: Pod
  metadata:
    labels:
        app: jos
        component: jobrequestservice
  spec:
    containers:

      -

        image: 'josappregistry.azurecr.io/
jobrequestservice:1.0'
        name: jobrequestservice
        ports:

          -

            containerPort: 3001
    imagePullSecrets:
    - name: josappsecret
```

If you compare the preceding `Pod` kind with the `Deployment` kind, you will notice that the Deployment contains the same attributes or keywords that the Pod configuration does. The reason we created the Deployment kind and not the Pod kind here is because we need to run multiple instances of Pod that can be configured using the `replicas` keyword. This assures us that the number of desired Pods instances will always be up and running.

`metadata` contains the required labels, while `spec` contains pod-specific attributes to configure. Here, we have containers where we specified the image that we deployed to the ACR, a `containerPort` where the application will run, and `imagePullSecrets`, which denotes the secret we created previously.

Labels and selectors are used to link different kinds of objects. So, in this document, we have a deployment kind and then a template that denotes Pod. As you may have noticed, we specified two labels, namely, app and `components`. Labels are just key-value pairs and can be of any key. However, the selector should reference the same key to establish linking between these two kinds of objects. So, labels such as app and component are defined under `template` and referenced as a selector in `deployment`. So, when the deployment and Pod objects are created in AKS, they will be linked to each other.

Before running the deployment, we need to create a secret on our machine that allows AKS to pull images from the ACR. To create a secret, execute the following command:

```
kubectl create secret docker-registry josappsecret --docker-
server=josappregistry.azurecr.io --docker-email=user@jos.com
--docker-username=josappregistry --docker-password=1tAchqYUdu5h
nZXuxPf2S=SEkaGTrC3
```

In the preceding command, specify the name of your secret. It should be the same as the one you specified in the deployment YAML file under the `imagePullSecrets` section or vice versa.

`--docker-server` is the complete address of the ACR, `--docker-email` could be any email address, and `--docker-username` and `--docker-password` can be obtained from the ACR **Access Keys** section, as shown in the following screenshot:

Figure 7.22 – The Access keys tab of the ACR

Open the document in VS Code and then use the **Kubernetes: Apply** option from VS Code's command palette:

Figure 7.23 – Applying the deployment to Kubernetes

This will apply the configuration and create two objects, called `deployments` and `pods`, respectively.

You can verify the deployments by running the following command:

```
kubectl get deployments
```

For the pod, execute the following command:

```
kubectl get pods
```

Once both deployments have been created, we won't be able to access them from the internet. To make them accessible, we need to create a service kind of deployment. Create a new service deployment file named `JobRequest_K8_Service.yaml` and use the following configuration:

```
apiVersion: v1
kind: Service
metadata:
  labels:
    app: jos
  name: jobrequestservice
spec:
  ports:
  - port: 3001
    targetPort: 3001
    protocol: TCP
  type: LoadBalancer
  selector:
    app: jos
    component: jobrequestservice
```

In the preceding file, we have a `Service` kind that contains two sections: `metadata` and `spec`. `metadata` contains the name of the service and the required labels. `spec` contains service-specific configuration properties, including `ports`, `type`, and `selector`.

The ports contain `port` and `targetPort` properties, where `port` can be any port that will be used by the consumer to access the service, and `targetPort` should be the same as what was specified for `containerPort` in the deployment file we created previously.

`selector` should point to the same labels that were specified in the deployment file that connects `Service` with the specified pods.

Open this file in VS Code and run the `Service` kind of deployment from the command palette, as we did previously. You can refer to the deployment files provided in this book's GitHub repository for other services.

Summary

In this chapter, we learned about containerizing multi-platform applications using Docker and deploying them to AKS. We started by setting up the Docker and Kubernetes CLIs in our development environment and learned about some concepts surrounding containers, along with their benefits. We learned about Docker as a technology used for containerizing applications and started building docker images for each microservice. After that, we explored various options available in VS Code that we can use to create and build Docker images. Then, we set up the ACR on Azure using VS Code and then AKS using the respective extensions. After that, we learned how to tag images and the commands that we can use to run to tag images and then push them to the ACR. Lastly, we understood the overall Kubernetes architecture and deployed images to AKS using VS Code.

In the next chapter, we will focus on using GitHub with VS Code for source control, continuous integration, and continuous deployment by using Azure Pipelines.

8
Git and Azure DevOps

In *Chapter 7, Deploying Applications on Azure*, we finished deploying the frontend and backend applications, and with that we completed the design, development, and deployment of our applications. In this chapter, we will look at using Git for version management and Azure for automating the deployment process.

In a real-life application, often several developers would work together to develop and test the application. They can work on different features of the application and can also be in different locations. This poses the challenge of how to manage the overall code base and how to share and synchronize code between developers and teams. Multiple version-management tools and techniques are available to facilitate this, such as **SVN** (short for **Subversion**), **TFVC** (short for **Team Foundation Version Control**), and Git. Our focus in this chapter will be to use Git from VS Code and create Git repositories in Azure DevOps.

The following topics will be covered in this chapter:

- Version-control overview

- Introduction to Git

- Creating repositories in Azure DevOps

- Using Git with VS Code

- Creating build pipelines and enabling Continuous Integration (CI) in Azure DevOps

- Creating release pipelines and enabling Continuous Delivery (CI) in Azure DevOps

Technical requirements

Before we start with Git, install Git from `https://git-scm.com/downloads`. The site will provide downloads for Windows, Mac OS, and Linux/Unix operating systems. Once downloaded, follow the wizard to install Git.

To check whether Git is installed and working, open a terminal window and run `git --version`. If Git is installed properly, you will see the installed Git version.

Version-control overview

Git is a version-control system: it tracks changes that are made to the source code. It is open source and is implemented by several vendors. Some of the Git hosting services are GitHub, Bitbucket, and Azure DevOps, among others.

Let's talk about what version control is and why it is important to have a system to track changes.

When we write code as a single developer or in a team, we start off by creating the basic structure and gradually move on and add features. The code base is continuously increasing and changing. Once you reach a particular milestone or complete a particular task, you will feel the need to create a backup or a snapshot of this moment. This ensures that if any future changes break the existing code, you will always have a backup to refer to. Secondly if there are multiple developers working on different parts of the code, at a certain point in time, it will be mandatory to synchronize and replicate the scattered code across the whole team. These changes can be spread over several folders and files, each file containing changes that were made in several different lines, whether different developers updated different files or one file was updated by several developers. Judging from these few scenarios, it almost seems impossible or extremely time consuming to carry out code synchronization using traditional copy-and-paste techniques. Now, multiply this concern by the number of times this activity has to be repeated. This is the reason version-management tools were created—to reduce the administrative overhead of efficiently maintaining the code base.

During the past years, multiple version-control systems have been created, such as VSS, SVN, and Git, and each have their own advantages and disadvantages. Our focus in this chapter is on Git, which we will explore in detail. SVN is also popular in developers, but compared to Git, one of the features where it lags behind is its central version-control system. The central server creates a single point of failure and will require network connection to commit changes. Git, on the other hand, works with a combination of local and remote repositories. This allows a developer to work offline and later synchronize changes with the remote server. If you are new to version control, then do spend some time reading about these other techniques as well. This will help you better understand why Git has a very high adoption rate and is popular in the developers community.

Introduction to Git

Before we start running Git commands, let's discuss the basic structure of Git. Git manages files in repositories. Repositories track changes to the files over a period of time and hold the history of what was changed, when it was changed, and who changed it. Files move between your working directory, staging area, local repository, and the remote repository. To understand this better, let's look at the overall flow of Git.

Git follows a specific flow to manage file versions locally and remotely. To maintain different versions, Git uses the concept of repositories. A repository is a folder inside your workspace and contains the suffix .git in its name. To understand how this folder is created and how versions are managed, let's look at the following figure:

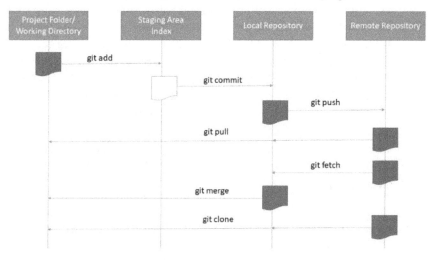

Figure 8.1 – Git flow

The preceding figure shows the overall Git flow; let's look at each step in detail using an example.

Initializing the Git repository

To start using Git, the first step is to initialize the Git repository.

Running git init in your project folder will create a .git folder in the project directory. This will hold the complete history of your project and is referred to as the **local repository**:

PC > Local Disk (C:) > My VS Code Projects > MyProject > .git >

Name	Date modified	Type	Size
hooks	7/22/2020 7:54 PM	File folder	
info	7/22/2020 7:54 PM	File folder	
objects	7/22/2020 7:54 PM	File folder	
refs	7/22/2020 7:54 PM	File folder	
config	7/22/2020 7:54 PM	File	1 KB
description	7/22/2020 7:54 PM	File	1 KB
HEAD	7/22/2020 7:54 PM	File	1 KB

Figure 8.2 – A .git folder, also called the local repository

Figure 8.2 shows the `.git` hidden folder created after running the `git init` command in the project folder.

> **Tip**
>
> Use the following commands to set the name and email for the author:
>
> `git config --global user.name "Your Name"`
>
> `git config --global user.email "your_email@domain.com"`

Staging files

Initializing a repository in your project folder will not tell Git to track changes. Only files added to the staging area are tracked. This is the next stage in the flow. To test this, let's add a dummy text file in the project directory.

Now, in the terminal window, run `git status`. As shown in following screenshot, Git has picked up the newly added file, but also shows that it is currently untracked:

```
C:\My VS Code Projects\MyProject>git status
On branch master

No commits yet

Untracked files:
  (use "git add <file>..." to include in what will be committed)

        File1.txt

nothing added to commit but untracked files present (use "git add" to track)
```

Figure 8.3 – Untracked files

Running `git add .` will stage all files. Alternatively, you can specifically add individual files to the staging area using `git add File1.txt`.

Committing files

The staged files are ready for a commit, so in the next stage, running `git commit -m "message"` will save the staged files in the local repository. Running `git log` will show the list of commits. Since we have only just done our first commit, you will only see one commit:

```
C:\My VS Code Projects\MyProject>git log
commit 4611c218b48ce95f4a28d314b5ed1bb430331488 (HEAD -> master)
Author: Khusro Habib <khusro.habib@gmail.com>
Date:   Wed Jul 22 20:38:24 2020 +0400

    our first commit
```

Figure 8.4 – Git logs

Figure 8.4 shows the unique `commit` hash, the `Author` who committed the change, and the timestamp when these changes were committed to the local repository. The latest commit is also marked as the `HEAD`.

Pushing changes to the remote repository

The local repository can serve the purpose of tracking changes locally, but when you are working with a team of developers, this is not enough. We use remote repositories to save changes and snapshots in a central location. This helps when sharing code between teams.

Let's create a remote repository on `dev.azure.com`. You can also use other providers, such as GitHub or Bitbucket:

1. Go to `https://dev.azure.com` and create a free account.

2. After logging in, create a new project by pressing the **New Project** button.

3. Enter the repository name and description, and choose whether you want to make your repository `public` or `private`. With `private`, you can specify who can see and commit to this repository.

Azure DevOps will create a project and also create one default repository with the same project name. To link our local repository to the newly created remote repository, run `git remote add <remote_repo_name> <repo_url>`. Look at the following screenshot:

```
C:\My VS Code Projects\MyProject>git remote add origin https://    @dev.azure.com/VSCodeBook/
LearnVSCode/_git/LearnVSCode
```

Figure 8.5 – Linking a local repository to a remote repository

Running `git remote -v` will show the remote repositories linked to the local repository. The local repository is now linked to the remote repository, but you will notice that the remote repository is still empty.

To push the complete history of changes to the remote repository, run `git push -u <remote_repo_name> <branch_name>`. To push the complete repository with all its branches, use `git push -u <remote_repo_name> --all`. Look at the following screenshot:

```
C:\My VS Code Projects\MyProject>git push -u origin --all
Counting objects: 3, done.
Writing objects: 100% (3/3), 236 bytes | 236.00 KiB/s, done.
Total 3 (delta 0), reused 0 (delta 0)
remote: Analyzing objects... (3/3) (4 ms)
remote: Storing packfile... done (254 ms)
remote: Storing index... done (79 ms)
remote: We noticed you're using an older version of Git. For the best experience, upgrade to a newer version.

To https://dev.azure.com/VSCodeBook/LearnVSCode/_git/LearnVSCode
 * [new branch]      master -> master
Branch 'master' set up to track remote branch 'master' from 'origin'.
```

Figure 8.6 – git push command to push local changes to the remote repository

After executing the `push` command, all changes will be pushed to the remote repository created on Azure DevOps. To check whether the local and remote repository are at the same level, run `git log` as follows:

```
C:\My VS Code Projects\MyProject>git log
commit 4611c218b48ce95f4a28d314b5ed1bb430331488 (HEAD -> master, origin/master)
Author: Khusro Habib <khusro.habib@gmail.com>
Date:   Wed Jul 22 20:38:24 2020 +0400

    our first commit
```

Figure 8.7 – Git log showing local HEAD and remote repository origins at the same commit level

Comparing *Figure 8.4* with *Figure 8.6*, you will notice the addition of `origin/master`. This means that the latest commit on the local repository specified by `HEAD->master` and the latest commit on the remote repository specified by `origin/master` are pointing to the same commit hash. If changes are made by other team members, you can fetch and merge them in your local repository using `git fetch` and `git merge`.

Alternatively, you can also use `git pull` to fetch and merge remote changes at the same time using one command. In the case that a remote repository already exists and you would like to start by creating a local repository and a working directory that replicates all changes, you can run `git clone <repo_url>`.

Having been introduced to Git and its key features, let's move ahead and explore Azure DevOps, which is used to manage our Job Order System project remote repositories.

Creating repositories in Azure DevOps

Following our introduction to version management and a basic idea of how Git works, let's get back to our project and see how we can use Azure DevOps to manage the project repositories.

Recalling our case study of Job Orders System, in total we have five projects. We have the frontend Angular project, a Java Spring Boot API project, a Node JS API project, a .NET Core project for managing interapplication messaging, and a Python API project for email messaging. Keeping in mind that later we want to create automated deployment pipelines, the best way to manage these projects in Azure DevOps is by creating separate repositories for each one of them.

We have already created a project in Azure DevOps for testing Git functionality. Let's use the same project and create the project repositories inside it:

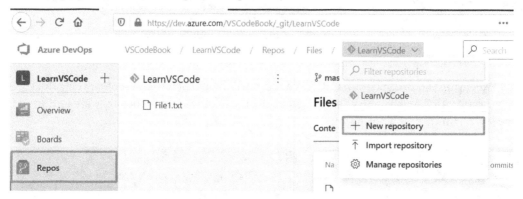

Figure 8.8 – Create a new repository in the Azure DevOps project

As shown in Figure 8.8, you should go to your **Azure DevOps** project and select **Repos** from the side bar.

Click the dropdown from the toolbar; you will see the previously created default **LearnVSCode** repository. This is the same repository that we used to explore the Git flow. Here, click on **New Repository**, leave the repository type as **Git**, enter the name of the repository, deselect **Add a README**, and click **Create**.

Similarly, create **JavaSpringBootAPI**, **NetCoreAPI**, **NodeJSAPI**, and **PythonAPI** repositories in Azure DevOps:

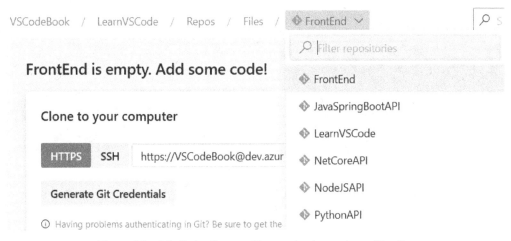

Figure 8.9 – Job Order System Git repositories on Azure DevOps

We can see that Figure 8.9 shows the list of repositories created for each project. All these project repositories are linked to one Azure DevOps project, **LearnVSCode**.

A project in Azure DevOps provides several other functionalities, such as managing work items, creating backlog, running sprints, and much more. Creating one Azure DevOps project for our Job Orders System will help us manage our complete product in one location.

With the remote repositories now ready, let's go back to our projects in VS Code and link them to their respective repositories in Azure DevOps.

Using Git with VS Code

Git is highly integrated in VS Code. The commands we explored a bit earlier in this chapter can all be executed from VS Code. To explore Git using VS Code, let's take the frontend project and explore the different Git functions from within VS Code.

Start by opening the frontend project folder and go to the **Source Control** tab by clicking on the following button:

Figure 8.10 – Button to reach Source Control

Our project folder does not contain a `.git` folder, which means that the first step is to initialize the repository. You can do this by either clicking the **Initialize Repository** button in the **Source Control** tab or by using the command palette and searching for `Git: Initialize Repository`:

Figure 8.11 – Options to initialize a Git repository from VS Code

Clicking on **Initialize Repository** will run `git init` for the current working directory.

Our project includes the `node_modules` folder. This folder contains the libraries that were downloaded using the node package manager. As a best practice, these libraries should not be tracked by Git, and VS Code will prompt you to add this folder to `.gitignore`. The `.gitignore` file is a file in which you specify the files or folders that should not be considered by Git when tracking changes. Choose **Yes** to exclude the `node_modules` folder from version tracking. On the bottom left, VS Code will show the default `master` branch:

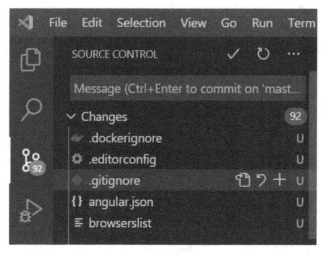

Figure 8.12 – Source Control tab showing untracked files

Figure 8.11 shows 92 files that can be added or staged for a commit in the Git repository.

The **U** sign beside each file shows that these are currently untracked by Git. Clicking the button with three dots in the top-right corner will show the set of Git commands that you can run on these files. Click on the three dots and select **Stage All Changes**. You will notice that the **U** sign will change to **A**. This means that the files have now been added for the next commit. This is similar to the **git add** command.

To commit the staged files to the local repository, select the **Commit All** options from the same menu. VS Code will prompt you to enter the commit message, so write a message and press *Enter*:

Figure 8.13 – Commit message in VS Code

All the stage changes will now be added to your local repository. Next, let's link our local repository with the remote repository **FrontEnd** on Azure DevOps:

1. To link the local repository with the remote repository, copy the repository URL from Azure DevOps. You can do this by selecting the **FrontEnd** repository from the drop-down menu in the toolbar and copying the URL from the **Clone to your computer** section.

2. Add the remote repository from VS Code by calling the command palette and selecting **Git Add Remote**:

Figure 8.14 – Link the local repository to the remote repository

3. Next, VS Code will prompt you to enter the URL. Paste the URL copied from the Azure DevOps site:

https://VSCodeBook@dev.azure.com/VSCodeBook/LearnVSCode/_git/FrontEnd

Add remote from URL https://VSCodeBook@dev.azure.com/VSCodeBook/LearnVSCode/_git/FrontEnd

Figure 8.15 – Remote repository URL

4. The remote repository should have a name; as a practice, we use `origin`.
    Enter `origin`:

Figure 8.16 – Enter the remote repository name

5. Here, VS Code will prompt you to enter the Git credentials. To get the credentials,
    click the **Generate Git Credentials** button in your repository on Azure DevOps.
    You will find this button on the **Clone Repository** page. Take the user ID, copy the
    generated password, and enter the password prompt.

    The remote repository will be added. You are now ready to push your changes to
    Azure DevOps.

6. Finally, select the source control action menu using the button with three dots
    and select **Push**. VS Code will ask if the master branch does not have an upstream
    branch. Select **Yes** and continue:

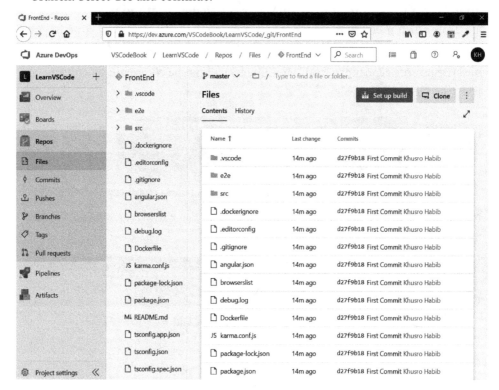

Figure 8.17 – Local repository changes pushed to the Azure DevOps FrontEnd repository

As you can see in Figure 8.17, all the files and commits created on the local repository are pushed to the remote repository. Select **Commits** from the navigation bar on the left and you will notice that the commit history is also synchronized.

By now, you must have noticed that VS Code provides a high level of integration for source control, and that, compared to the command-line option, it's pretty simple and user friendly. Next, let's look at a very important feature of Git called branching, and how VS Code supports it.

Branching and merging

Another important feature of the Git version-control system is the option to create branches.

While working on software products, we often come across scenarios where we want to keep our production code base intact and then create a separate copy for adding features for handling bugs. A typical way of handling this is to create a copy of your entire code and then start making changes there. Once the changes are complete, we merge the copied code and the original version together. Using branches, we can do the same in an efficient manner.

In our previous examples, you must have seen the word `master` branch. This is the default branch that is created when initializing a repository. Adding a branch will create a parallel version of the same code that can then be changed and later merged with the `master` branch. There are several different practices when using branches. A branch can be created to fix bugs, add a feature, or even manage different versions.

There are several ways in which developers and teams use branching and merging. It provides great flexibility in managing changes that are made by several team members to several different topics. To give you an overview of how developers use this feature, next up we will look at some widely used practices.

Some practices of using branches

One way to use branches is to have the master branch reflect your production release. A second branch, such as `development` or any other name that suits you, can reflect the changes that will be released in the next stable version. Nightly builds can also be created from the development branch. This can help you keep the production stable version intact in case any issues arise in the stable version while new features are developed in parallel.

In the case of bug fixes, a branch can be created off the master branch where all necessary changes are made. Once these changes are ready for deployment, a new production version is created and the bug-fix branch is merged into the master branch. You can use `Git Tag` to mark the version on your master branch. Later, the bug-fix branch should be merged into the development branch as well, so that the recent bug fixes are available in the development branch as well. After this, the bug-fix branch can be deleted.

If different team members are working on different features, it might be a good idea to create separate branches for each feature from the development branch. This way, whenever a feature is ready, it will be merged back into the development branch and will be included in the next production release. This will allow different teams to work on different topics at varying paces and will also provide an opportunity to discard a feature if it is not worth keeping.

If you have a common code base that is replicated to create multiple long-lived product versions that will also run in parallel for a long time, then one option could be to branch off `version2018`, `version2019`, `version2020`, and so on from the master branch. Here, each `versionXXXX` will have its own development branch. The `versionXXXX` branch will be treated as the master branch for that version release and each release to that version will be tagged on its relevant `versionXXXX` branch.

This was an introduction to branching and some ways in which it can be used to efficiently manage your code base.

Next, let's look at a basic workflow of creating a feature branch on our frontend using VS Code and Azure DevOps.

Branching in VS Code and Azure DevOps

To create a new branch in VS Code, click on the master branch in the bottom-left corner. The command palette will show an option to **Create new branch**. Select this and create a branch from the currently selected branch. You can also use **Create new branch** to use another branch. Next, enter a name of the branch, we'll use **feature1,** and press *Enter*. The new branch is created and selected as well. In the bottom-left corner, you will see the active branch is **feature1**. Change the title in **index.html**, **Commit** the changes, and **Push** to the remote repository. VS Code will prompt you to publish the new branch since no upstream branch exists. Click **OK** to continue:

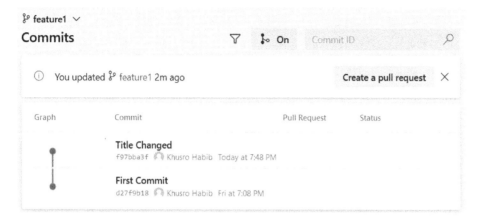

Figure 8.18 – feature1 branch pushed to Azure DevOps with commit history

Let's assume that your changes are complete and ready to be merged with the master branch. Instead of directly merging the feature branch into your master branch, in a real-world scenario, the new code will go through a review process before it is merged with the master branch. To handle the review process in Azure DevOps, we will create a pull request; this can be enforced by setting a branch policy. You can do this by clicking the **Create a pull request** button, as shown in Figure 8.19.

While creating the pull request, you will add a title and description for the reviewer. Select one or more reviewers. You can also link it to a particular **Work Item** of this project. Clicking **Create** will create a pull request and trigger an email to be sent to the approver. The approver can compare the changes and either **Approve**, **Approve with suggestions**, **Wait for author**, or **Reject the request**:

Figure 8.19 – Pull request pending for approval

Once the request is approved, you can click **Complete** to merge the `feature1` branch into `master`:

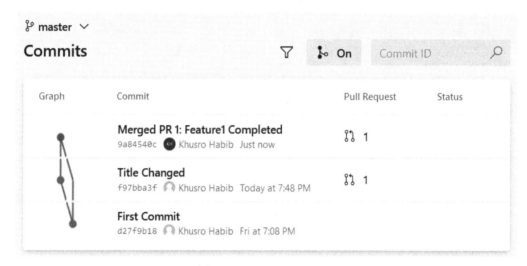

Figure 8.20 – Commit history for master branch post pull request completion

As shown in Figure 8.20, you can see that the commit history of the **feature1** branch has been updated in the master branch and that all changes have been merged together. We also selected the option to delete the `feature1` branch after completion.

> **Tip**
>
> To revert or reset committed changes, use the following commands:
>
> `git revert <commit_hash>`: Reverts the changes in the specified commit by creating a new commit.
>
> `git reset --hard <commit_hash>`: Resets the branch to the specified commit.
>
> `git reset --keep <commit_hash>`: Resets and brings changes to the working directory.

With this, we wrap up our discussion on version control and how it can be used to effectively manage source code. In the next section, we will look at ways of automating the build and deployment processes using CI/CD.

Automating build and deployment using CI/CD

Today, the IT industry as a whole is going through a major transition. IT is changing from being a support function to a strategic capability. The traditional business models are under continuous threat of disruption, and this is forcing organizations to make heavy investments in adopting new technology, as well as exploring innovative ideas. Product development is more focused on ideating, trying, deploying, and checking results compared to executing several months of planning, development, and deployment cycles. Agility is the key in winning out over competition, and this has forced IT companies to think and work on methods to remove bottlenecks that slow down the delivery of value to customers.

IT as a whole has made tremendous progress, and has been addressing these concerns. Some of them involve using cloud technology, which speeds up the delivery of infrastructure while reducing administrative overheads. Agile product-management methodologies and the availability of software to support them, such as Azure DevOps, speeds the delivery of product changes by automating build and deploy cycles and also streamlines communications between development and operations. All of these advancements have made it possible for organizations to ideate, develop, and deploy quickly to compete with the market and stay ahead of their competition.

Continuous integration (**CI**) and **continuous delivery** (**CD**) focus on the same principles and allows a wide team of developers and product owners to manage and deliver changes faster and more efficiently. With version control as the base for code management, CI helps in speeding up build cycles and provides an easy way to automate build processes that are triggered by code merges. On the other hand, CD streamlines the release cycles by automating the process of taking the successfully built code, testing it using the user acceptance test, and then releasing it to production.

Both of these techniques are very well supported by Azure DevOps, and in this section, we will look at how we can automate the build and deployment process by using CI/CD pipelines.

Creating a CI pipeline in Azure DevOps

Now that we have looked at the background and benefits of automating the build and deployment steps, let's go through an example and create a pipeline for the NodeJSAPI project:

1. The repository for the Node project is already created in Azure DevOps. Click on the **Pipelines** link to create our first build pipeline.

2. On the Pipelines page, click the **Create Pipeline** button. Here, you can select the repository in which your code is located.

3. Since we are using Azure DevOps for version control, select **Azure Repos Git** and then select the **NodeJSAPI** repository.

4. In the **Configure** section, select **Docker (Build a Docker image)**. The wizard will prompt you to specify the location of your Dockerfile. Our file is located in the root folder of the repository, so we click **Validate** and configure it. Azure DevOps has generated the yml file; you can rename it by clicking on the name:

Figure 8.21 – Creating or editing a CI pipeline

Figure 8.21 shows the screen where you can create and edit your CI pipeline. On the left side is the yml editor. On the right side, you can search for predefined **Tasks** and add them to the yml file on the left. The **Variables** button in the toolbar allows you to define user-defined variables.

Let's discuss the yml file in detail. It starts with a `trigger` that defines the action that will automatically run the pipeline. You can specify the branches that will receive the trigger. In our generated yml file, Azure DevOps has the default master branch:

```
trigger:
- master
```

The `resources` keyword defines `self`; this refers to the repository where our yml file is stored:

```
resources:
- repo: self
```

A `pool` defines where the job is run. You can select the vmImage on which you want the build steps to run. It's also possible to configure and use a local agent pool by using `pool: default`:

```
pool:
  vmImage: 'ubuntu-latest'
```

After specifying the trigger and pool, you can specify the steps that you want to run. These can be as simple as running a bash command or even a predefined task from Azure DevOps. It is also possible to group the set of steps into a `job` and group multiple jobs under `jobs`. Jobs can be grouped into stages, and can also be used to define dependencies.

We did not require stages in our pipeline, however, for the sake of this example, we have added one stage:

```
stages:
- stage: Build
  displayName: Build image
```

Under the `stage`, we have created one job and specified the required steps there. In reality, if you have only one job, then you can even skip the `jobs` and `job` addition. In this case, your steps will be bundled into a job implicitly:

```
jobs:
- job: Build
  displayName: Build and Save Image
  pool:
    vmImage: 'ubuntu-latest'
```

When defining a build pipeline, you will often require dynamic data to be passed to tasks or commands. The pipeline provides you with an option to declare variables, and as well as user-defined variables, Azure DevOps provides you with a list of predefined variables. You can view the list of predefined variables at `https://docs.microsoft.com/en-us/azure/devops/pipelines/build/variables?view=azure-devops&tabs=yaml`.

Finally, we define the steps to be executed. The first step is a predefined Docker task in which we build the image. The name mentioned in the `displayName` parameter will appear during execution. Then we specify the Docker `build` command and the path to the `dockerfile`. When defining the *Dockerfile* path, we use a predefined variable, `Build.SourcesDirectory`. This refers to the agent's directory path, where our source code files will be downloaded from the repository.

In the arguments, we pass the `-t` option to tag our image. `ContainerRegistryName` is a variable that we created to specify `josappregistry.azurecr.io`; this is our Azure Container Register name. Next, we created the `NodeJSAPIName` variable and initialized it to `jobrequestservice`; this refers to the Node JS API hosted on AKS. When combined, these variables will represent the Docker image name. `Build.BuildId` is a predefined variable that refers to the ID of the completed build; we add this to create a new Docker image version each time a build is run:

```
steps:
- task: Docker@2
  displayName: Build an image
  inputs:
    command: build
    dockerfile: '$(Build.SourcesDirectory)/Dockerfile'
    arguments: '-t $(ContainerRegistryName)
/$(NodeJSAPIName):$(Build.BuildId)'
```

In the next step, we again call the predefined `Docker@2` task; however, this time, we call it with the `Save` command. We save the image as a `tar` file using the `-o` addition and specify the directory that the artifact will be stored on the agent using a predefined variable, `Build.ArtifactStagingDirectory`. We use the `NodeJSAPIName` user variable to name the `tar` file and `$(ContainerRegistryName)/$(NodeJSAPIName):$(Build.BuildId)` to refer to the recently saved image:

```
- task: Docker@2
    displayName: Save Image
    inputs:
      command: save
      arguments: '-o $(Build.
ArtifactStagingDirectory)/$(NodeJSAPIName).
tar $(ContainerRegistryName)/$(NodeJSAPIName):$(Build.BuildId)'
```

In the last step, we publish the `tar` file as an artifact so that it can be picked up by the release pipeline for deployment.

Here, we use the predefined task `PublishBuildArtifacts@1`. `PathtoPublish` refers to the directory in which we saved the Docker image as a `tar` file. `ArtifactName` specifies the name of the artifact and `publishLocation` specifies that the artifact should be stored in the Azure pipeline container:

```
- task: PublishBuildArtifacts@1
    inputs:
      PathtoPublish: '$(Build.ArtifactStagingDirectory)'
      ArtifactName: drop
      publishLocation: Container
```

The CI pipeline is ready. Click **Save** and run to execute it:

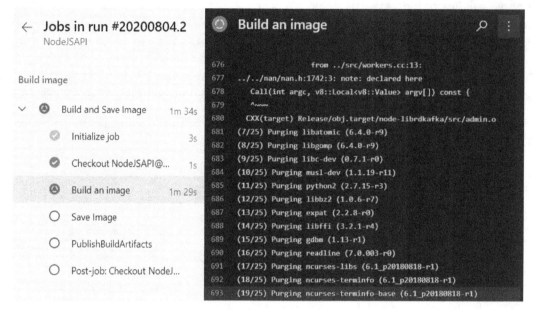

Figure 8.22 – Job execution

Figure 8.22 shows the job execution, with each step executing one after the other.

The CI pipeline is set. Let's go and create the CD release pipeline to deploy the saved image to `Azure Kubernetes Services`.

Creating a CD pipeline in Azure DevOps

To manage the automated deployment of our build, we will use the Azure DevOps release pipelines. In the pipeline menu, select **Releases** and create a new release pipeline. Azure DevOps will prompt you to select a template; in our case, we will start with an **Empty Job**.

If you recall, we completed the build steps in our CI pipeline by publishing the Docker image `tar` file as an artifact. Click on `Add an artifact` and select **Build**; here, we are adding the latest published drop artifact:

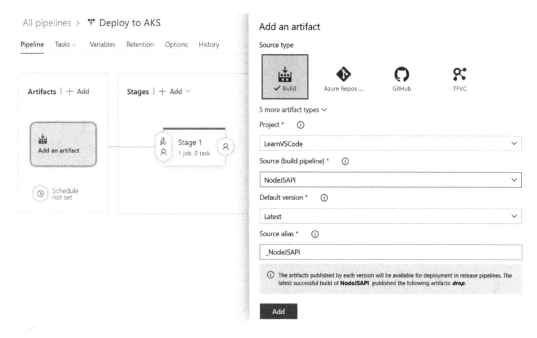

Figure 8.23 – Adding an artifact in the CI pipeline

In Figure 8.23, we configure the settings so that the release pipeline will pick the artifact from the latest build run. In the **Project**, we selected **LearnVSCode**, and in the **Source**, we specify the name of our CI build pipeline. The **Default version** refers to the build run and lastly, the alias.

The message shows that our last successful build created an artifact with the name **drop**; this will be picked by the release pipeline. Click **Add** and complete the artifact selection. A release pipeline can be triggered manually or automatically based on an event. You can select a trigger by pressing the trigger button in the top-left corner of the artifact tile:

Figure 8.24 – The trigger button

In *Figure 8.24*, we specify * in the **Build branch filters** tab; this directs Azure DevOps to trigger a release whenever new a build is run on any of the branches. It's possible to enter multiple conditions that are evaluated using an OR operation. If any one condition matches, a release will be created:

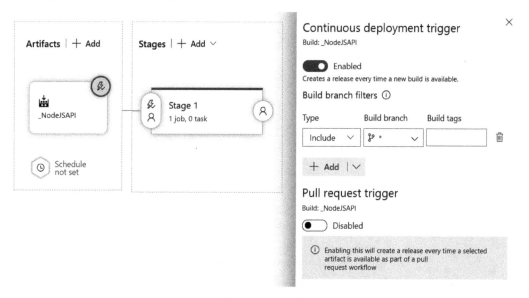

Figure 8.25 – Configuring triggers for a release pipeline

In a typical scenario, before a build is deployed to production, you will have a test environment where user acceptance testing scenarios will be run. Once this has been completed, your build will be deployed to production. You can use stages for these scenarios.

Before configuring the stages, you should define variables in the release pipeline variables section, as shown in the following figure:

Name	Value	🔒	Scope
ContainerRegistryName	josappregistry.azurecr.io		Release
k8jobreqdeployment	jobrequestservice		Release
NodeJSAPIName	jobrequestservice		Release

Figure 8.26 – User-defined variables for the CD release pipeline

Let's see how we can do this:

- Click on the link under **Stage 1** to configure the release steps. In **Stage Tasks**, click on **Agent Job** to configure the agent settings. In the agent specifications, select **ubuntu-20.04**. Next, we will add the tasks required to load the artifact and deploy the image on **AKS**.

- Click the plus sign and search for the **Docker** task. Click on the task and change its name to `load`.

- In the command field, enter `load`, and in the arguments field, enter the following command:

```
--input $(System.DefaultWorkingDirectory)/_NodeJSAPI/
drop/$(NodeJSAPIName).tar
```

This task will run the `docker load` command to pick the Docker image `tar` file created during the build phase.

- In the next step, we configure the `docker push` command:

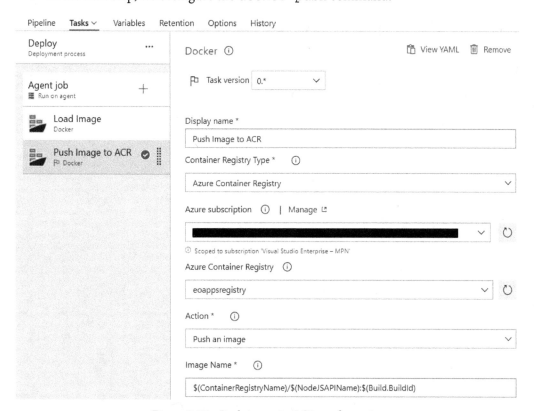

Figure 8.27 – Push image to ACR configuration

In Figure 8.27, we configure the task to push the loaded image to the `Azure Container Registry`. In the settings, select the **Azure Subscription**, the container registry, and the image name. We refer to the image name using the following defined variables:

```
$(ContainerRegistryName)/$(NodeJSAPIName):$(Build.BuildId)
```

In the last step, we add a task to set the latest published image in our AKS cluster.

To do this, add a new task for `kubectl`. Once the `kubectl` task is added in the pipeline, go to the settings pane of the `kubectl` task and select `Kubernetes Service Connection` as the service connection type.

If you don't already have a service connection to your AKS cluster, click `New` and add a connection. In the `command` field, specify `set`, and in the arguments field, specify the following command:

```
image deployment/$(k8jobreqdeployment)
$(k8jobreqdeployment)=$(ContainerRegistryName)
/$(NodeJSAPIName):$(Build.BuildId)
```

In this command, we are telling the AKS cluster to point the job request service deployment to the newly pushed image on the ACR.

We use the user-defined variables declared earlier in Figure 8.26. Finally, click `Save` and click on the **Create Release** button to trigger the release pipeline:

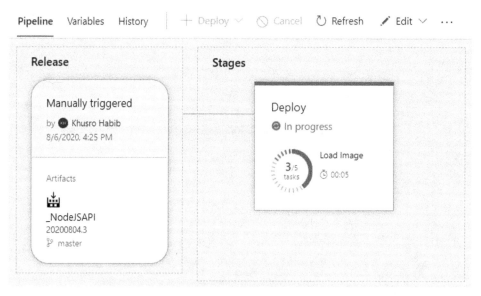

Figure 8.28 – Execution of the CD release pipeline

Figure 8.28 shows the release pipeline under execution. You can drill down for further details. With this, we have completed and covered the automation of build and deployment processes using Azure DevOps.

Summary

In this chapter, we looked at why version management is essential and explored Git as a version-control system.

To have an overview of Git, we ran an end-to-end Git flow using VS Code. We also learned about the importance of branching, discussed some of the best practices, and later on, explored branching with an example. At the end of this chapter, we learned how continuous integration/continuous delivery can work together seamlessly with version control and automate the build and deployment processes.

In the next chapter, we will look at extensions in more detail and create our own custom extension.

Section 3: Advanced Topics on Visual Studio Code

This section of the book covers advanced topics, which include building custom extensions, their setup environment, and just how important remote development is in Visual Studio Code.

This section comprises the following chapters:

- *Chapter 9, Creating Custom Extensions in Visual Studio Code*
- *Chapter 10, Remote Development in Visual Studio Code*

9
Creating Custom Extensions in Visual Studio Code

In previous chapters, we have looked at different aspects of VS Code. Our focus throughout this book has been to take specific use cases and experience how VS Code can be used to address a particular requirement. There are some features available out of the box and others that are adapted to extend a certain tooling capability. VS Code, with its extension framework, has created an ecosystem where it's possible for contributors to enhance tooling features for their teams and also make it available for the wider community to benefit.

So far, we have explored extensions in terms of usage and looked at different scenarios where they can help speed up and enhance our development experience. However, despite all that is available in the extension marketplace, there will always be requirements that mean we have to create new extensions.

For this reason, our focus in this chapter will be to use the VS Code extension framework and create different types of extensions.

In this chapter, we will be covering the following topics:

- Setting up the environment for creating extensions
- Creating a new extension using TypeScript
- Creating a Kubernetes objects extension
- Creating a theme extension

So let's get started and look at the prerequisites for developing extensions for VS Code.

Technical requirements

Before we start creating custom extensions we will need to install Yeoman and VS Code's Extension Generator. This will allow us to generate the scaffolding for the extension project. To install Yeoman and `generator-code` in one go, run the following command:

```
npm install -g yo generator-code
```

The preceding command will install the Node packages globally. To check if the installation was successful, run `yo -version` and `yo code` in the terminal.

With `yeoman` and the VS Code Extension Generator set up, let's look at how we can create different types of extensions.

Creating a new extension

When creating a custom extension there are several templates available. To build an extension from scratch we will be using the `yeoman` and VS Code Extension Generator. The Extension Generator provides different options to quickly scaffold the extension project. To explore it in detail, let's create our first extension using the **New Extension** template.

Generating the extension project

The VS Code Extension Generator provides a quick and easy way to scaffold your extension project. Let's work through the following steps to create the project:

1. Start off by running the following command in a terminal window:

    ```
    yo code
    ```

    This is how the output appears when you run the command:

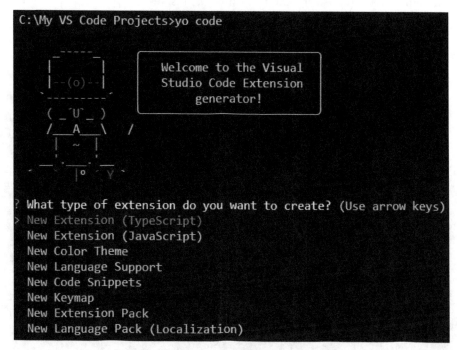

    Figure 9.1 – Yeoman Extension Code Generator

The preceding screenshot shows a prompt where the Extension Generator is asking you to select the type of extension you would like to create. Select **New Extension (TypeScript)** and press *Enter*. There is another variant with *JavaScript* in case you would like to use that.

Then enter the following details.

*   The **Name** refers to the display name of your extension:

    Name of the extension: VS Code Book First Extension

*   The **Identifier** refers to the internal name of your extension. This can be specified as a word or multiple words with dashes. By default, the generator will suggest a display name with dashes – you can use that or choose to specify your own:

    Identifier: vscode-first

- The **Description** is more of a short text to explain what the extension is for:

Description: This is our First VS Code Book extension using typescript

- To initialize a local repository enter Y, which means **Yes**:

Initialize a git repository: *Enter (Y)*

Select your preferred package manager, npm or yarn:**Package Manager:** npm

This is what you get as the output:

```
? What type of extension do you want to create? New Extension (TypeScript)
? What's the name of your extension? VS Code Book First Extension
? What's the identifier of your extension? vscode-first
? What's the description of your extension? This is our First VS Code Book extension using type script
? Initialize a git repository? Yes
? Which package manager to use? npm
  create vscode-first\.vscode\extensions.json
  create vscode-first\.vscode\launch.json
  create vscode-first\.vscode\settings.json
  create vscode-first\.vscode\tasks.json
  create vscode-first\src\test\runTest.ts
  create vscode-first\src\test\suite\extension.test.ts
  create vscode-first\src\test\suite\index.ts
  create vscode-first\.vscodeignore
  create vscode-first\.gitignore
  create vscode-first\README.md
  create vscode-first\CHANGELOG.md
  create vscode-first\vsc-extension-quickstart.md
  create vscode-first\tsconfig.json
  create vscode-first\src\extension.ts
  create vscode-first\package.json
  create vscode-first\.eslintrc.json

I'm all done. Running npm install for you to install the required dependencies.
If this fails, try running the command yourself.
```

Figure 9.2 – Generating a new extension using Yeoman Extension Generator

The preceding screenshot shows you the Extension Generator creating the extension project and running npm install to download the required packages. With this, the project has been generated, so you can now open the vscode-first folder in VS Code.

After this, we will move on to understanding extension configuration.

Discussing the extension configuration

The project scaffolding has now been generated and we are ready to start further development of our extension. Before we add some features, let's discuss the extension configuration specified in the `package.json` file. We will leave the script and dependencies section since that is generic to npm, and will instead focus on the parameters specific to the extension project.

The following three parameters refer to the entries made while generating the project:

```
'name': 'vscode-first',
'displayName': 'VS Code Book First Extension',
'description': 'This is our First VS Code Book extension
using type script',
```

Moving on, `version` refers to the extension version, `engines` specifies the minimum VS Code version the extension is compatible with, and `categories` defines the extension category:

```
'version': '0.0.1',
'engines': {
    'vscode': '^1.48.0'
},
'categories': [
    'Other'
],
```

For further details, refer to the VS Code Extension Manifest documentation at the following link:

https://code.visualstudio.com/api/references/extension-manifest

Moving further, `activationEvents` specifies the event that will activate your extension. There are several types of events, the details of which can be found at the following the link:

https://code.visualstudio.com/api/references/activation-events

In `activationEvents`, we are using the `onCommand` event. This will activate the extension when `helloWorld` is called from the command palette:

```
'activationEvents': [
    'onCommand:vscode-first.helloWorld'
],
```

Moving on, the `main` parameter specifies the entry point of our extension. We have generated the *TypeScript* version of the extension, so we will be coding in the `extension.ts` file, whereas the `package.json` file refers to the transpiled version `extension.js`:

```
'main': './out/extension.js',
```

Finally, the `contributes` keyword specifies the command that, when called, will trigger the activation event `onCommand: vscode-first.helloWorld`:

```
'contributes': {
    'commands': [
        {
            'command': 'vscode-first.helloWorld',
            'title': 'Hello World'
        }
    ]
},
```

For further details on `contributes` refer to the following link:

https://code.visualstudio.com/api/references/contribution-points#contributes.commands

With our introduction to the `package.json` file complete, let's dive into the `extension.ts` file where the extension is coded.

Developing the extension

In the `extension.ts` file, you will see two methods, `activate` and `deactivate`. The `activate` method is called when the activation event specified in the `package.json` file is fired.

Let's now look at the `activate` method.

The first time that the `helloWorld` command is called from the command palette, VS Code will trigger the `activate` method and print **Congratulations, your extension 'vscode-first' is now active**. With this it will also execute the anonymous function specified with `vscode.commands.registerCommand` and show an information message reading **Hello World from VS Code Book First Extension!**. Refer to the following code:

```
export function activate(context: vscode.
ExtensionContext) {

    console.
log('Congratulations, your extension 'vscode-
first' is now active!');
```

From then on, whenever the `helloWorld` command is triggered from the command palette, VS Code will only execute the following code block, skipping the `console.log` message:

```
    let disposable = vscode.commands.
registerCommand('vscode-first.helloWorld', () => {
        vscode.window.
showInformationMessage('Hello World from VS Code
                           Book First Extension!');
    });
    context.subscriptions.push(disposable);
}
```

The extension is ready. Press *F5* to open another instance of VS Code for testing. Just like any other program, you can use breakpoints to debug the extension. It's a nice way to check how many times each method is called when the `Hello World` command is executed from the command palette.

In the newly opened instance of VS Code, call the command pallet and search for `Hello World`. As shown in the following screenshot, the command from your extension shows up:

Figure 9.3 – Executing the Hello World command

Hitting the command for the first time will call the `activate` method. Since we have placed a breakpoint in our `extension.ts` file, the following screenshot shows the `activate` method being called:

```
 7  export function activate(context: vscode.ExtensionContext) {
 8
 9      console.log('Congratulations, your extension "vscode-first" is now active!');
10
```

Figure 9.4 – The activate method on calling the extension for the first time

After the code in the activate method is executed, the code block specified with the `vscode.commands.registerCommand` is executed. The following screenshot shows the breakpoint stopping on the code block:

```
11      let disposable = vscode.commands.registerCommand('vscode-first.helloWorld', () => {
12          vscode.window.
13          showInformationMessage('Hello World from VS Code Book First Extension!');
14      });
15
```

Figure 9.5 – Code block specified with the register command being called

The preceding code will show an information message in the newly opened VS Code instance, as shown in following screenshot:

Figure 9.6 – Information message shown in VS Code

The basic generated extension project is running, and this should have given you an idea of the extension framework. Let's look at some additional options and also add few more commands to our project.

Categorizing commands

Often, your extension will have several commands, and it's easier for the user if all commands are grouped under a relevant category. To do this, we will add the `category` parameter for our `Hello World` command in the `package.json` file. The following code shows the category `Learn VS code` added to the `helloWorld` command:

```
{
    'command': 'vscode-first.helloWorld',
    'title': 'Hello World',
    'category': 'Learn VS Code'
}
```

Run the extension again and search for your command in the command palette. This time, search for `learn`. The following screenshot shows how your command can now be searched for with a category:

Figure 9.7 – Searching for commands by category

Similarly, we can register multiple commands to define different actions for our extension. Let's look at how we can add more commands to our extension. We will also look at how you can organize your extension project.

Adding the Open Folder command

Just like any other development project, you can arrange your code base into multiple files for easier management and future changes. As of now, the complete code resides in the `extension.ts` file and as the extension features are added, it will become difficult to maintain the whole project in one file. To deal with this, let's look at how you can organize your project into separate files.

Let's add one more command to open a folder. As we discussed earlier in the chapter, instead of adding this command in the same `extension.ts` file, we will create a separate `commands.ts` file and refer it in the main `extension.ts` file.

In the `/src` folder, create a `Commands` folder and add a new file called `commands.ts`. To register a new command, we need to provide the command name and the function to be executed. In the newly added `commands.ts` file, add the necessary code by working through the following steps:

- Start by importing the `vscode` package:

```
import * as vscode from 'vscode';
```

- Next, create a constant holding the command name:

```
export const myOpenFolderCommand = 'vscode-first.
myOpenFolder';
```

In the following code, we create a `JavaScript` function and call the `vscode.commands.executeCommand` method to open the file explorer:

```
export async function myOpenFolderFunc() {
    vscode.commands.executeCommand('vscode.openFolder');
}
```

- In the extension.ts file, we will import the commands.ts file by using the following code:

```
import * as mycommands from './Commands/commands';
```

- In the activate method, register the command and the JavaScript function. Here we refer to the extension command name and the function code from the commands.ts file:

```
//Adding the Open Folder command
  context.subscriptions.push(
    vscode.commands.registerCommand(
      mycommands.myOpenFolderCommand,
      mycommands.myOpenFolderFunc
    )
  );
```

In the package.json file, add the newly added command to the contributes section:

```
{
            'command': 'vscode-first.myOpenFolder',
            'title': 'Open Folder',
            'category': 'Learn VS Code'
}
```

Since both the extension-specific commands and their code are not registered until the activate method is called, we will specify the new command in activationEvents as well. If this is not done, then user will be forced to execute the helloWorld command before executing any other command:

```
'activationEvents': [
    'onCommand:vscode-first.helloWorld',
    'onCommand:vscode-first.myOpenFolder'
],
```

We are now ready, let's give it a try. Launch the extension by pressing *F5* and call the command pallet to search for Open Folder:

Figure 9.8 – Calling the Open Folder command

This will launch the file explorer where the user can select the folder to be opened in VS Code. Next, let's see how user input takes place.

Taking user input

The VS Code extension framework provides rich features to develop a user-friendly extension. As you may have noticed in other extensions, you can integrate the VS Code user input feature in your own extension. To explore this feature, let's add another command and use the showInputBox option. Let's see how this is done:

- Start by adding a new extension command in the commands.ts file:

```
export const myuserInputCommand = 'vscode-first.
userInput';
```

- Next, we create the function to be executed when our extension command is called from the command palette. In the following code, we request the user for input using the vscode.window.showInputBox command:

```
export async function myUserInputFunc() {
  const userInput = await vscode.window.showInputBox({
    placeHolder: 'Please enter message',
    prompt: 'Enter Message',
  });
```

The entered message is displayed using the vscode.window. showInformationMessage command as follows:

```
  if (userInput !== undefined) {
    vscode.window.showInformationMessage(userInput);
  }
}
```

- Next, we add the newly added command to the activate method:

```
//Adding the user input command
context.subscriptions.push(
  vscode.commands.registerCommand(
    mycommands.myuserInputCommand,
    mycommands.myUserInputFunc
  )
);
```

- Finally, add the newly added `userInput` extension command to `activationEvents` in the `package.json` file:

```
'activationEvents': [
    'onCommand:vscode-first.helloWorld',
    'onCommand:vscode-first.myOpenFolder',
    'onCommand:vscode-first.userInput'
],
```

- Also, add the extension in the `contributes` section, as follows:

```
{
    'command': 'vscode-first.userInput',
    'title': 'User Message',
    'category': 'Learn VS Code'
}
```

The command is now ready, so let's give it a try. Launch the extension using *F5* and call the command palette, searching for `User Message`:

Figure 9.9 – User Message command

As shown in the following screenshot, pressing *Enter* will show an input box where the user can enter their message:

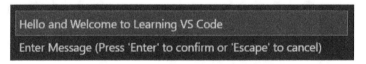

Figure 9.10 – User input in VS Code

Hitting *Enter* will show the message, as demonstrated in the following screenshot:

Figure 9.11 – User message being displayed

For further details on the set of commands available, you can visit the official documentation at `https://code.visualstudio.com/api/references/commands` and also further explore using **File | Preferences | Keyboard Shortcuts**.

With this, we have completed our first extension using `TypeScript` and also looked at some features of the overall VS Code extension framework. Next, we will explore the code snippet extension and develop our Kubernetes objects extension.

Creating the Kubernetes objects extension

Kubernetes (K8s) is a very demanding topic when working with microservices or containers. We have seen in the previous chapters, how you can declaratively deploy K8s objects such as `Deployments` and `Services` inside K8s clusters, for which we create a YAML file containing the K8s object-specific information. The YAML file has a fixed-tab delimited format and a single space or invalid indentation can cause the deployment to fail. Let's see how this happens.

Generating the code snippet extension project for K8s objects

In this section, we will create an extension that will help us to scaffold the `Deployment` and `Service` templates for K8s objects. Let's begin:

1. To start with, we will first create a new folder named `K8sext`. Then run the following command to start building the extension:

    ```
    yo code
    ```

2. From the list of extensions, choose `New Code Snippets` shown as follows:

```
? What type of extension do you want to create?
  New Extension (TypeScript)
  New Extension (JavaScript)
  New Color Theme
  New Language Support
> New Code Snippets
  New Keymap
  New Extension Pack
  New Language Pack (Localization)
```

Figure 9.12 – Choose New Code Snippets extension type

3.   Enter the **name** of the extension as `k8sext`, the **identifier** of the extension as `k8sobj`, add a **description**, and for **language** ID, specify `YAML`, as follows:

```
? What type of extension do you want to create? New Code Snippets
Folder location that contains Text Mate (.tmSnippet) and Sublime snippets (.sublime-snippet)
or press ENTER to start with a new snippet file.
? Folder name for import or none for new:
? What's the name of your extension? k8sext
? What's the identifier of your extension? k8sobj
? What's the description of your extension? Kubernetes Objects Scaffolding Extension
Enter the language for which the snippets should appear. The id is an identifier and is single,
lower-case name such as 'php', 'javascript'
? Language id: yaml
    create k8sobj\.vscode\launch.json
    create k8sobj\package.json
    create k8sobj\vsc-extension-quickstart.md
    create k8sobj\README.md
    create k8sobj\CHANGELOG.md
    create k8sobj\snippets\snippets.code-snippets
    create k8sobj\.vscodeignore

Your extension k8sobj has been created!

To start editing with Visual Studio Code, use the following commands:

    cd k8sobj
    code .

Open vsc-extension-quickstart.md inside the new extension for further instructions
on how to modify, test and publish your extension.

For more information, also visit http://code.visualstudio.com and follow us @code.
```

Figure 9.13 – Creating the K8s Code Snippet Extension

A new code snippet extension template will be generated in the `K8sext` folder. We can now open the folder in VS Code to further customize the basic extension template.

4.   Modify the version, change it to 1.0.0, and also add a publisher of VS Code
     Book:

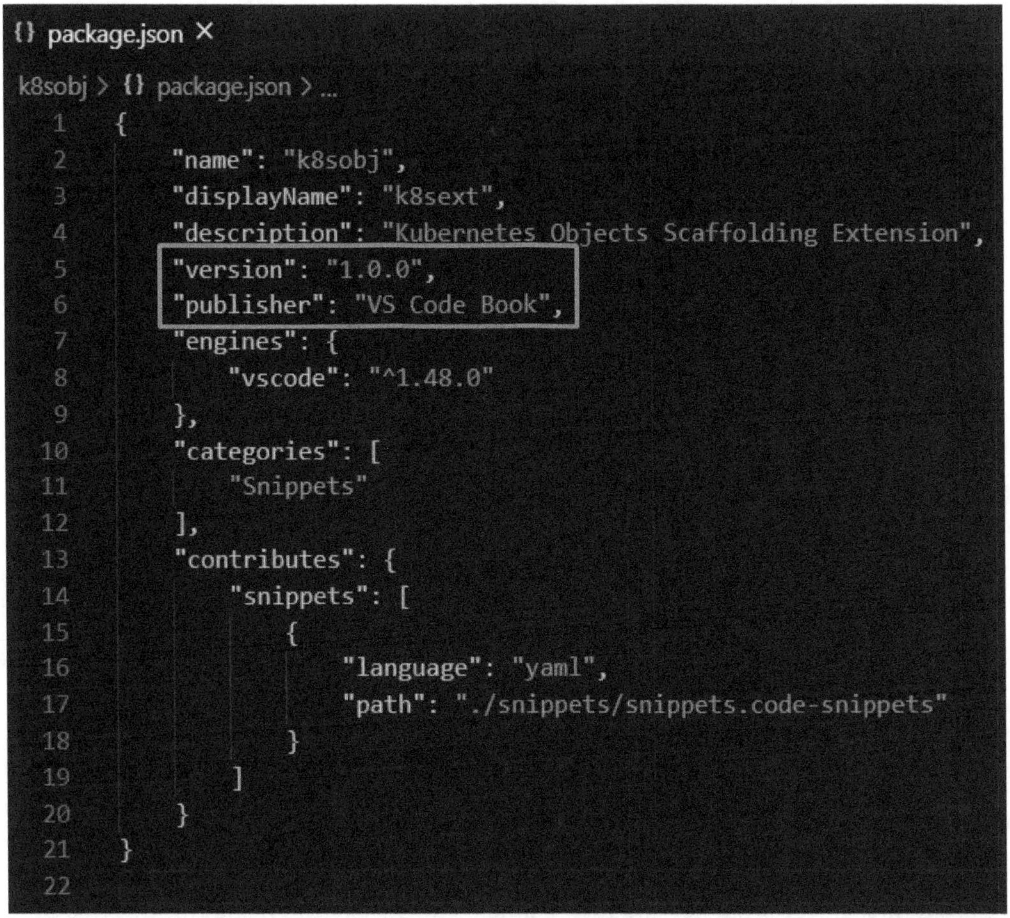

```
{} package.json ✕

k8sobj > {} package.json > ...
  1    {
  2        "name": "k8sobj",
  3        "displayName": "k8sext",
  4        "description": "Kubernetes Objects Scaffolding Extension",
  5        "version": "1.0.0",
  6        "publisher": "VS Code Book",
  7        "engines": {
  8            "vscode": "^1.48.0"
  9        },
 10        "categories": [
 11            "Snippets"
 12        ],
 13        "contributes": {
 14            "snippets": [
 15                {
 16                    "language": "yaml",
 17                    "path": "./snippets/snippets.code-snippets"
 18                }
 19            ]
 20        }
 21    }
 22
```

Figure 9.14 – Modifying version and publisher information

5.   Next, expand the `snippets` folder and open the `snippets.code-snippets` file as shown in the following screenshot:

Figure 9.15 – Selecting the snippets file

This is the file where we will add the code snippet to scaffold K8s objects. Now that we've covered this, let's look at how to update the extensions.

Creating the code snippet for K8s Deployment and Service objects

With the code snippet extension project generated and the required configurations updated, let's look at the different sections of the `snippets.code-snippets` file and also update it to create the code snippet extension.

We need to first create a root key and then define the `prefix`, `body`, and `description` sections of the extension. `prefix` is the actual command we execute inside the **YAML** file to scaffold the template, the `body` essentially contains the code snippet (in our case, the K8s `Deployment` or `Service` objects), and `description` contains some more information about the extension:

```
{
   'Kubernetes Deployment Object Scaffolding Extension': {
      'prefix': '!k8deploy',
      'body': [
      ],
      'description': 'Kubernetes Deployment Extension'
```

```
    }
}
```

In the preceding code snippet, the `prefix` to call this extension is `!k8deploy`. Now, let's add the K8s `Deployment` template in the `body` array. In certain cases, the template may contain some placeholders that need to be changed as per the scope or requirements. We can define a placeholder in our code snippet using the following syntax:

```
${IndexNo: Value}
```

`IndexNo` is the tab index that defines the order in which the cursor will move between the placeholders in sequential order.

Here is the complete JSON for the extension, containing the K8s `Deployment` object inside `body` and you can see that it also contains some placeholders that allow the user to define their values when scaffolding the template:

```
{
    'Kubernetes Deployment Object Scaffolding Extension': {
        'prefix': '!k8deploy',
        'body': [
            'apiVersion: apps/v1',
            'kind: Deployment',
            'metadata:',
            'name: ${1:deploymentname}',
            'spec:',
            'replicas: ${2:1}',
            'selector:',
            '    matchLabels:',
            '        app: ${3:appname}',
            '        component: ${4:componentname}',
            'template:',
            '    metadata:',
            '    labels:',
            '        app: ${3:appname}',
            '        component: ${4:componentname}',
            '    spec:',
            '    containers:',
            '        - ',
```

```
'          image: \'${5:imagename}\'',
'          name: ${1:deploymentname}',
'          ports:',
'            -',
'              containerPort: ${6:80}',
'    imagePullSecrets:',
'    - name: ${7:imagepullsecretname}'
  ],
  'description': 'Kubernetes Deployment Extension'
}
}
```

We can test this extension by copying the extension folder inside the .vscode\ extensions directory. In Windows, you can find the .vscode folder under the Users directory.

Before proceeding, let's add the K8s Service template in the same snippets. code-snippets file. We will add the code snippet for the K8s Service after the K8s Deployment code snippet covered in the preceding paragraphs. Here is the code snippet for the Service extension:

```
'Kubernetes Service Object Scaffolding Extension': {
   'prefix': '!k8service',
   'body': [
     'apiVersion: v1',
     'kind: Service',
     'metadata:',
     '  labels:',
     '    app: ${1:appname}',
     '  name: ${2:servicename}',
     'spec:',
     '  ports:',
     '  - port: ${3:80}',
     '    targetPort: ${4:80}',
     '    protocol: TCP',
     '  type: ${5:LoadBalancer}',
     '  selector:',
```

```
    '        app: ${1:appname}',
    '        component: ${6:componentname}'
    ],
    'description': 'Kubernetes Service Extension'
}
```

`!k8service` can be used to scaffold the K8s `Service` template, and there are some placeholders that can be replaced with actual values at the time of scaffolding.

To test this extension, we need to copy the whole extension folder and place it inside the `.vscode/extensions` folder. Once it is done, open VS Code and create a new workspace or open a new folder.

Verify that the extension is listed in the installed `Extensions` list shown as follows:

Figure 9.16 – Selecting the custom-built k8sext extension

To test these extensions now, we will create two files, namely `samplek8sdeploy.yaml` and `samplek8sservice.yaml`. In the `samplek8sdeploy.yaml` file, we will scaffold the K8s `Deployment` template and in the `samplek8sservice.yaml` file, we will scaffold the K8s `Service` template. Refer to the following screenshot:

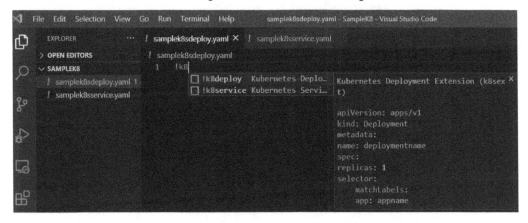

Figure 9.17 – Scaffolding snippet for Deployment extension

On pressing *Enter*, it will scaffold the template for the K8s `Deployment` object:

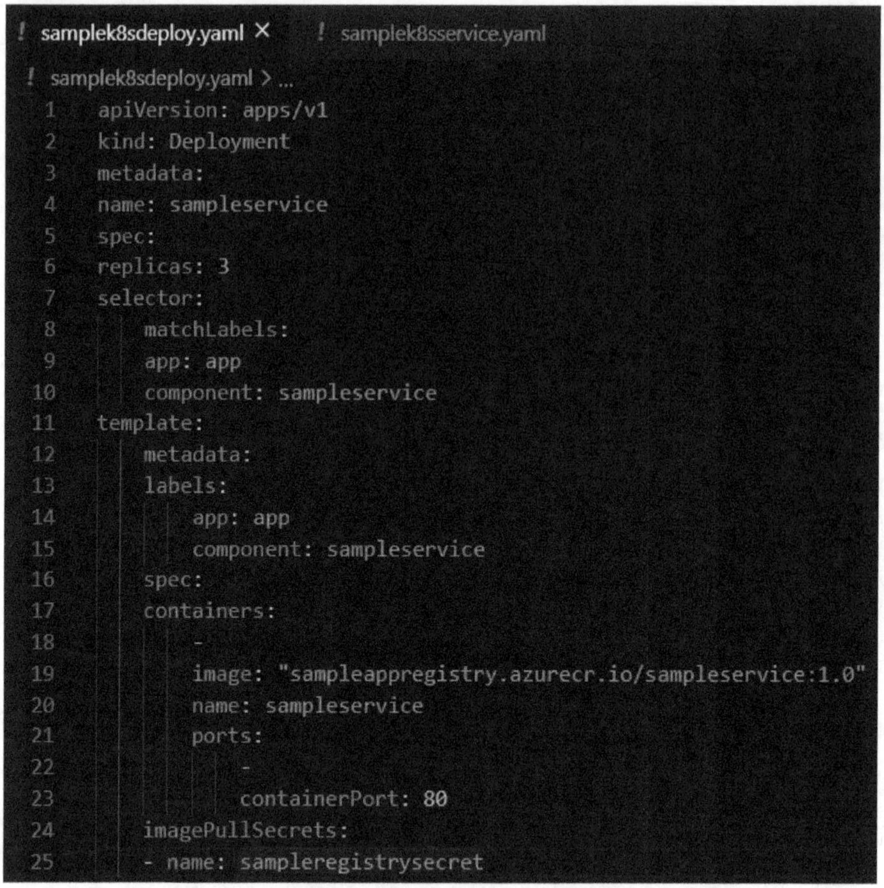

```
samplek8sdeploy.yaml ×        samplek8sservice.yaml

 samplek8sdeploy.yaml > ...
  1    apiVersion: apps/v1
  2    kind: Deployment
  3    metadata:
  4    name: sampleservice
  5    spec:
  6    replicas: 3
  7    selector:
  8        matchLabels:
  9        app: app
 10        component: sampleservice
 11    template:
 12        metadata:
 13        labels:
 14            app: app
 15            component: sampleservice
 16        spec:
 17        containers:
 18            -
 19            image: "sampleappregistry.azurecr.io/sampleservice:1.0"
 20            name: sampleservice
 21            ports:
 22                -
 23                containerPort: 80
 24        imagePullSecrets:
 25        - name: sampleregistrysecret
```

Figure 9.18 – Configuring values for K8s Deployment object

To scaffold the template for the K8s `Service` object, we select the second option and it generates the template as follows:

```
 !  samplek8sdeploy.yaml          !  samplek8sservice.yaml  ✕

 !  samplek8sservice.yaml > {} spec > {} selector > 🔤 component
 1     apiVersion: v1
 2     kind: Service
 3     metadata:
 4       labels:
 5         app: app
 6       name: sampleservice
 7     spec:
 8       ports:
 9       - port: 80
10         targetPort: 80
11         protocol: TCP
12       type: LoadBalancer
13       selector:
14         app: app
15         component: sampleservice
```

Figure 9.19 – Configuring values for K8s Service object

With this code snippet extension type, you can easily build your own extensions and accelerate development. Next, we will explore the theme extension and look at how we can create a custom theme.

Creating the theme extension

Theme extensions are useful to colorize your code based on your own color scheme. It's very easy to build and scaffold a standard configuration that can be customized to your requirements. Let's see how to go about this.

Generating the theme extension project

To create a new theme extension, you can run the same yo code command and select New Color Theme. Refer to the following screenshot:

```
C:\Books\VSCode\extension\themeext>yo code
Unable to fetch latest vscode version: Error: [object Object]


    _-----_
   |       |
   |--(o)--|        Welcome to the Visual
   `---------`      Studio Code Extension
    ( _`U`_ )           generator!
   /___A___\   /
    |  ~  |
  __`.___.'__
`-         -`

? What type of extension do you want to create? New Color Theme
? Do you want to import or convert an existing TextMate color theme? (Use arrow keys)
> No, start fresh
  Yes, import an existing theme but keep it as tmTheme file.
  Yes, import an existing theme and inline it in the Visual Studio Code color theme file.
```

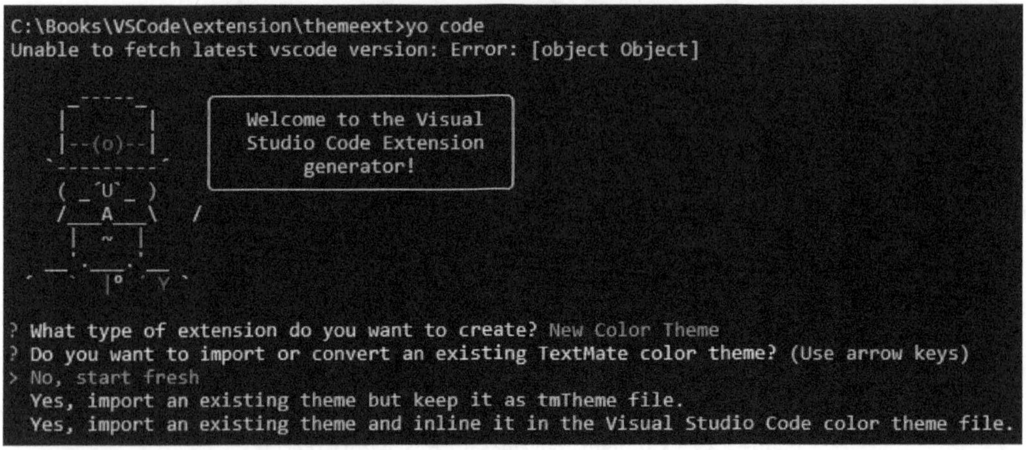

Figure 9.20 – Creating a New Color Theme extension

When you opt to create a theme extension, it will ask whether to create a new one from scratch, import an existing theme file, or add the VS Code color theme file itself. Furthermore, provide a name, identifier, description, and display name, and specify the base theme, choosing from Dark, Light, and High Contrast. Refer to the following screenshot:

```
? What's the name of your extension? booktheme
? What's the identifier of your extension? booktheme
? What's the description of your extension? Color Theme Extension
? What's the name of your theme shown to the user? booktheme
? Select a base theme: Dark
   create booktheme\themes\booktheme-color-theme.json
   create booktheme\.vscode\launch.json
   create booktheme\package.json
   create booktheme\vsc-extension-quickstart.md
   create booktheme\README.md
   create booktheme\CHANGELOG.md
   create booktheme\.vscodeignore
```

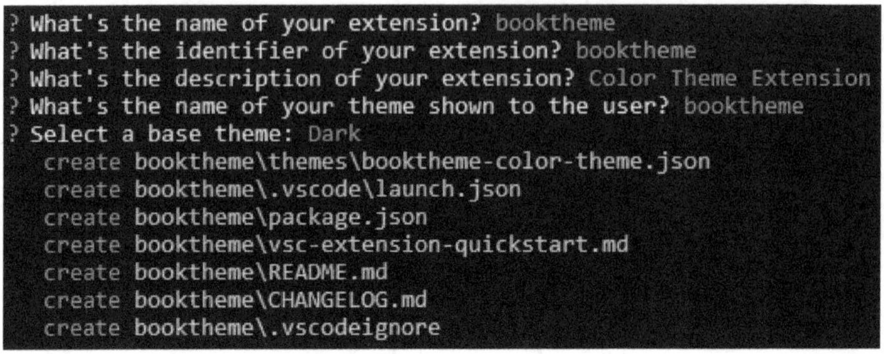

Figure 9.21 – Entering metadata about the new extension

We select **Dark** as the base theme and then open the folder using VS Code for further customization. The following screenshot shows the folder structure created when generating the new theme extension project:

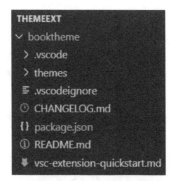

Figure 9.22 – Folder structure of new theme extension

Modify the package.json file, set the **version** as 1.0.0, and add the **publisher** property to define the publisher's name. With the theme extension project generated and the configuration updated, let's look at creating our custom theme.

Creating the theme

The theme details can be modified by opening the booktheme-color-theme.json file, as shown in the following screenshot:

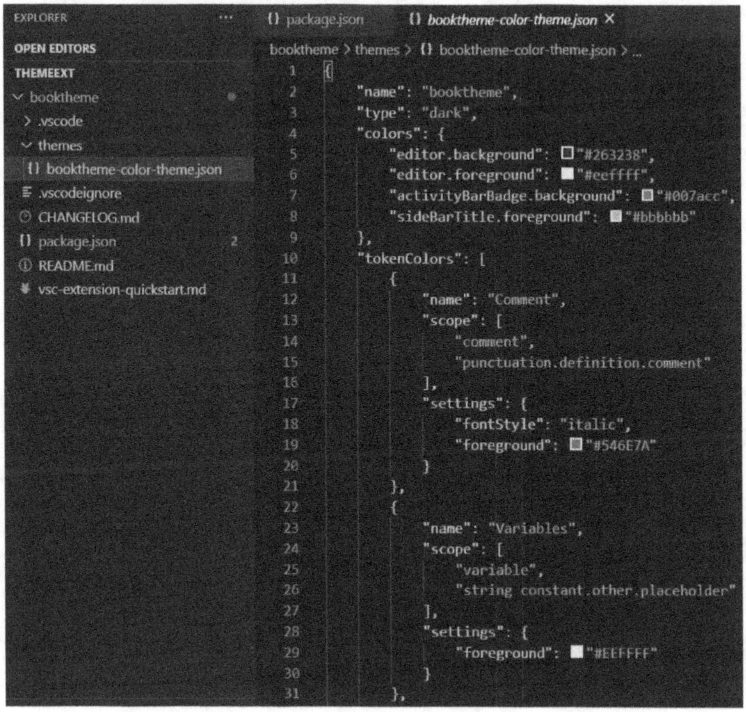

Figure 9.23 – Theme extension configurable values

For each section, you can modify the `background` and `foreground` colors. The `colors` section holds the editor's background and foreground color properties and the `tokenColors` section contains the default background and foreground color values for different fields. You can modify the field colors as per your requirements in the `settings` section.

Let's modify the `background` and `foreground` colors of the editor and also modify the `fontStyle` and `foreground` color for the `Comment` element as follows:

```
booktheme > themes > {} booktheme-color-theme.json > [ ] tokenColors > {} 0 > {} settin
 1    {
 2        "name": "booktheme",
 3        "type": "dark",
 4        "colors": {
 5            "editor.background": □"#012356",
 6            "editor.foreground": ■"#ffffff",
 7            "activityBarBadge.background": ■"#007acc",
 8            "sideBarTitle.foreground": ■"#bbbbbb"
 9        },
10        "tokenColors": [
11            {
12                "name": "Comment",
13                "scope": [
14                    "comment",
15                    "punctuation.definition.comment"
16                ],
17                "settings": {
18                    "fontStyle": "bold underline",
19                    "foreground": ■"#d6f014"
20        }
```

Figure 9.24 – Editing some fields

Save the preceding modification and copy it in the VS Code extension folder on your machine. After copying the folder, it should appear inside the installed extensions:

Figure 9.25 – Setting a color theme for the newly created extension

To test the extension, let's create a new file and use the colors that we configured in the code for the extension:

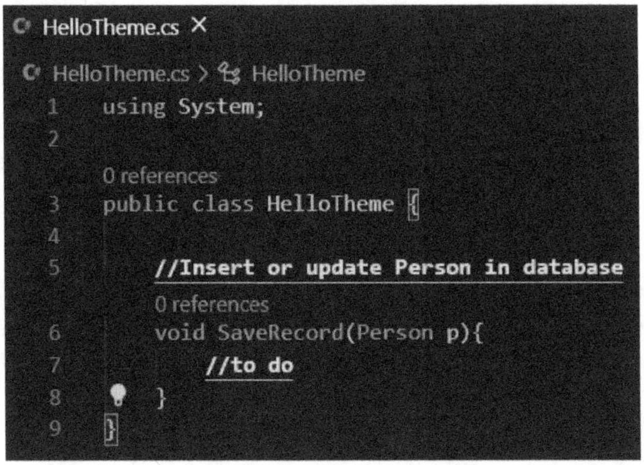

Figure 9.26 – Using the extension

In the preceding sample `HelloTheme.cs` file, notice that the font style and color for comments (//) are the same as we configured in our theme file.

Summary

In this chapter, we learned how to create custom extensions in VS Code. We started off by setting up the environment and moved on to creating our first extension using `TypeScript`. We then created a code snippet extension for the K8s objects, which will come in handy since the `YAML` file for the K8s deployment is very specific in terms of indentations. Finally, we created a simple theme extension and explored how it can be used to change the theme of your VS code environment.

You are now able to understand the extension framework in detail and have gained the knowledge necessary to create your own extensions. You have also learned how the VS Code Extension Generator along with Yeoman   can be used to quickly scaffold your project. You can also now differentiate between the different types of extension projects and know how they can be developed to provide various productivity features.

In the next chapter, we will conclude this book by exploring remote development and its different use cases.

10
Remote Development in Visual Studio Code

Remote development is an essential feature of VS Code. It allows you to develop and build applications in a remote environment that can be any machine on premises or in the cloud. There are several scenarios where developers need to develop, modify, or debug code on the remote machine to make it work. If you have a team of developers working remotely in different locations, there can be scenarios where you want to support your team members to debug or modify the application code. The standard practice is you ask the team member to share their screen using any of the tools available and help them fix or write the code side by side. With remote development in VS Code, you no longer need to access the remote machine directly. Instead, you can connect to the remote environment from your machine where VS Code is running and access the environment where all the dependencies reside, along with the application code, and help fix the problem.

Moreover, if you are a part-time developer for one of the projects, you don't need to set up the whole development environment on your machine. Instead, you can use the remote development feature in VS Code to connect to the remote machine and do the development there.

VS Code offers a seamless experience for remote development, where there is nothing that would suggest to the developer that the application is not residing locally and accesses the full capabilities of VS Code to perform remote development. Remote development allows developers to edit, build, or debug application code on a remote environment, whether it's a container or a machine, running on any OS such as Windows, Linux, or macOS. After installing specific remote extensions, you will be able to connect to the remote environment and develop, build, debug, and troubleshoot any application using all the capabilities of VS Code.

In this chapter, we will be covering the following topics:

- Understanding the basic architecture of how remote development works
- Setting up remote development in VS Code
- Developing application on macOS using the Remote – SSH extension
- Developing application inside a Docker container using the Remote – Containers extension hosted inside a container
- Developing an application inside an Ubuntu machine using the Remote – WSL extension
- A brief introduction to GitHub Codespaces

Understanding remote development in VS Code

VS Code is one of the few editors that supports remote development. When we say *remote development*, that means you only need to install VS Code on your local machine and after setting up remote development extensions, you can access the remote environment that contains the code, application dependencies, and other components to do the development.

Some scenarios where VS Code remote development works well include the following:

- Suppose you are a Windows developer. Your application can run cross-platform and the operating system you desire to run your application on is macOS. Using remote development, you can connect to a remote macOS environment and work from your Windows PC using VS Code.
- When you don't have space or cannot install more components or frameworks that are necessary to build the application locally.

- When your application requires lots of dependencies and configuration takes a lot of time. With remote development, the contributors to the application can directly build the application on the remote machine.

- If you don't have the necessary tools and frameworks to build the application locally.

- When you need to develop a Linux deployed application using the **Windows Subsystem for Linux (WSL)**.

- If you are a part-time developer and the frequency of contribution is low. Rather than setting up everything locally, you can use the remote development feature and contribute.

- When you want to debug or troubleshoot applications running on a remote machine.

- When you have applications running inside containers and you want access to the actual folder and to be able to update or modify files.

Once the remote development extensions are set up in VS Code, you can use VS Code to execute commands on the remote machine. While connected with the remote environment, any extension you install will be installed on the remote machine and once disconnected, those extensions will not be available locally. Being a developer, you get a similar experience as you would have when working on a local environment.

Let's have a brief look at how remote development works:

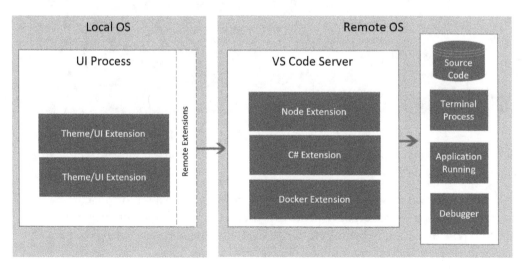

Figure 10.1 – Remote development architecture

In the preceding diagram, starting from the left side, it shows VS Code installed on a local machine.

With remote extensions from the local machine, you can connect to the remote machine. Once the connection is established, you can install and run extensions on the remote machine to build, debug, and run applications depending on the scenario. The application source code, terminal process, application, and debugger—all these components are a part of the remote machine. Developers can execute and run the extensions from VS Code, and it runs on the remote server. For example, you are working on a Node.js application. Once connected to the remote machine, the command running locally from VS Code will be executed on the remote machine.

Now that we understand what this is, let's see how to set up a remote environment in VS code.

Setting up remote development in VS Code

To set up remote development, you need to install the remote development extensions on a local machine. The remote extension pack contains extensions for SSH, containers, and the WSL.

To start with installing the remote extension pack, open the **Extension** tab, search for `remote development extension pack` and click on **Install**:

Figure 10.2 – Remote development extension pack

Once the extension is installed, you can connect to any folder on a remote machine, in a container, or in the WSL.

The **Remote – SSH** extension is used to connect to any folder on a remote machine, whereas the **Remote – Containers** extension is used to connect to a container and the **Remote – WSL** extension to connect to the WSL. These extensions can also be separately installed as well if you don't want to use all of them.

The following screenshot has a list of extensions that can be installed separately:

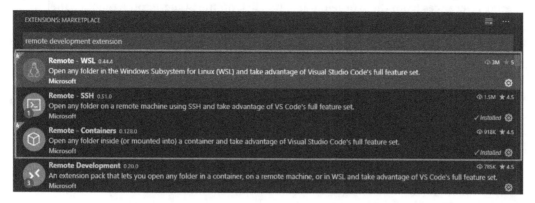

Figure 10.3 – Remote development extensions

Once the extensions are installed, you can do remote development through VS Code.

Let's now understand these extensions in detail in the following sections.

Using the Remote – SSH extension

The **Remote – SSH** extension allows you to open any remote folder on a remote machine where an SSH server is running. The target server should have SSH installed and the local OS should have an OpenSSH-compatible SSH client installed.

To see the system requirements for SSH for both local and remote machines, refer to the following link:

https://code.visualstudio.com/docs/remote/ssh

In this section, we will set up and install the SSH client on the Windows 10 OS and also set up the SSH server on macOS, then connect to the macOS machine through VS Code's **Remote – SSH** extension to modify application code.

To install the OpenSSH client on Windows, run the following PowerShell command:

```
Add-WindowsCapability -Online -Name OpenSSH.Client*
```

The preceding command needs elevated permissions to not throw an error, so run this command from PowerShell while running the latter as an administrator.

To configure an SSH server on macOS, we need to enable **Remote Login**. To do so, open **System Preferences** from the **Apple** menu, and click on the **Sharing** preferences panel as shown:

Figure 10.4 – System preferences in macOS

To verify if this feature is running properly, execute the following command on your macOS:

```
ssh localhost
```

On running this command, you should see the following output:

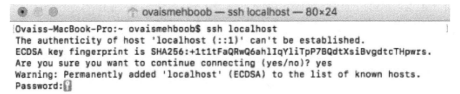

Figure 10.5 – Executing SSH command

As the basic setup is now done, we can use the **Remote – SSH** extension from VS Code to access the code stored on macOS. VS Code, under the hood, creates an SSH tunnel and executes the commands on a remote machine that are initiated from the local machine.

Now let's connect to the SSH server using the VS Code Remote – SSH extension and create a simple .NET Core application:

1. Open VS Code from the local machine and in the command palette, enter `remote` and select **Remote-SSH: Connect to Host…**:

Figure 10.6 – Use Remote-SSH extension to Connect to Host

2. Once you select the command, click on **Add New SSH Host..** and enter the host address. You can get this information from the **Sharing** preferences panel in macOS as shown in *Figure 10.4*. Enter it as follows:

Figure 10.7 – Entering the SSH command to connect to the remote machine running macOS

3. Next, select **macOS**:

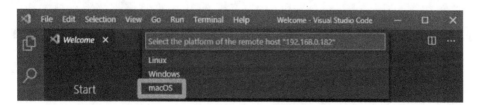

Figure 10.8 – Selecting macOS to connect using the Remote-SSH extension

4. Click on **Continue** to proceed with the default fingerprint option:

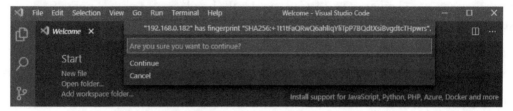

Figure 10.9 – Hit Continue to proceed to the next step

5. Enter your macOS password:

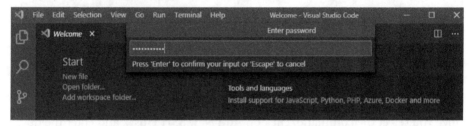

Figure 10.10 – Entering the macOS password

Once the credentials are verified, it establishes a connection to the remote machine and downloads some dependencies on the local machine:

Figure 10.11 – Downloading dependencies and setting up the connection

6. Once it is configured, you can see it connected with the SSH server running on macOS:

Figure 10.12 – Showing the IP once the connection is established

7. Once this is done, open the folder from VS Code, and it will show the remote folders of the macOS machine. Here we will select the `dotnetapp` folder created in macOS and create a new .NET Core console application:

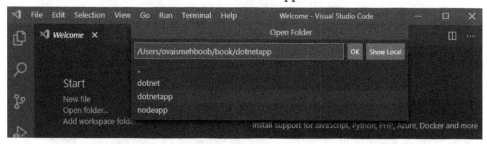

Figure 10.13 – Selecting the dotnetapp folder in macOS to connect using Remote-SSH extension

Finally, it opens up the macOS folder as shown here:

Figure 10.14 – Showing a folder on remote machine

To spin up a new .NET Core application, open the terminal window in VS Code and execute the following command:

```
dotnet new console
```

To build the .NET Core application, use the following command:

```
dotnet build
```

Run the application using the following command:

```
dotnet run
```

Here is the output in the terminal window that displays a **Hello World** message when the application is run:

```
PROBLEMS   OUTPUT   DEBUG CONSOLE   TERMINAL

Ovaiss-MacBook-Pro:dotnetapp ovaismehboob$ dotnet build
Microsoft (R) Build Engine version 16.6.0+5ff7b0c9e for .NET Core
Copyright (C) Microsoft Corporation. All rights reserved.

  Determining projects to restore...
  All projects are up-to-date for restore.
  dotnetapp -> /Users/ovaismehboob/book/dotnetapp/bin/Debug/netcoreapp3.1/dotnetapp.dll

Build succeeded.
    0 Warning(s)
    0 Error(s)

Time Elapsed 00:00:02.20
Ovaiss-MacBook-Pro:dotnetapp ovaismehboob$ dotnet run
Hello World!
Ovaiss-MacBook-Pro:dotnetapp ovaismehboob$ []
```

Figure 10.15 – Running a sample .NET Core console application

In this section, we used the **Remote – SSH** extension to successfully connect to a remote machine running macOS and build the simple .NET Core application to test remote development.

Using the Remote – Containers extension

The **Remote – Containers** extension allows you to develop, build, and run an application inside a container. You can build and run your application in a sandbox environment without having any direct effect on the hosting system.

To work with a container, we need Docker to be installed on the hosting machine. You can use Docker as a technology to build containers, mount a volume, and use VS Code to connect to that mounted folder.

> **Note**
>
> To learn more about the minimum system requirements for working with containers in VS Code, refer to the following link:
>
> `https://code.visualstudio.com/docs/remote/containers`

You can type `Remote-Containers` into the comand pallette to list multiple options to choose from. We will use the **Remote-Containers: Open Folder in Container...** option and build a simple application inside it.

To start with, let's open VS Code:

1. Start by selecting the **Remote – Containers** from the command palette:

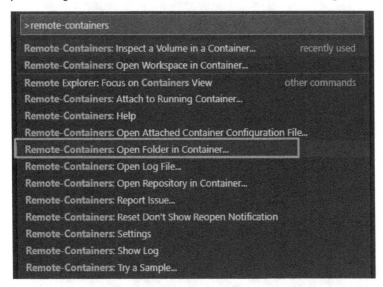

Figure 10.16 – Selecting the Remote-Containers: Open Folder in Container option

On selecting the highlighted option of the preceding screenshot, it will ask you to select the local folder that you want to mount to the container instance. Here you can create any folder (in our case we created a folder named `Container-NodeJS`).

You need to specify the complete path to this folder.

2. Next, it will ask you to select the language or platform to build the application. Here, you can select **Node.js 14**:

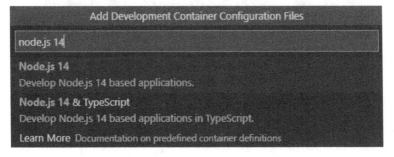

Figure 10.17 – Selecting Node.js 14 to develop a Node application

After selecting it, the container starts provisioning and the files will be created on the local machine:

Figure 10.18 – The image is provisioned on a container

3. After setting up the dev container, a root folder is created called `.devcontainer`. It contains two files, namely, `devcontainer.json` and `Dockerfile`:

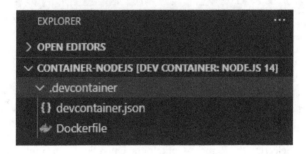

Figure 10.19 – Folder structure

Next, let's create a new Node.js Express app inside a container:

4. To do so, we will use the terminal window in VS Code. In the terminal window, you will notice that the folder is actually mounted inside the container, as shown here:

Figure 10.20 – Terminal window shows the container drive folder path

5. Run the following command to install the Node.js Express application generator inside the container:

```
npm install -g express-generator
```

6. Now, run the `express` command to scaffold a basic Node.js Express app project inside the container. You will see the files being created and you can modify them from the **EXPLORER** tab in VS Code:

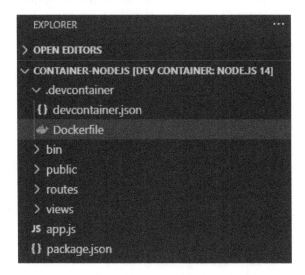

Figure 10.21 – The Node.js Express app files

We will now install the Node.js dependencies by running `npm install`, build the application using `npm build`, and run the application using `npm start`.

When building an application inside a container, the extensions installed on the hosting system will not be available. Hence, you need to install the respective extensions as needed on the container.

> **Note**
>
> On adding more extensions, you will get a warning about the extension running outside your container. To work with extensions, you need to switch to containers. Do this and it will reload your VS Code and then whatever extension you execute will run inside your container.

VS Code provides multiple ways to do remote development, in which developing inside a container is just one option, as we have learned previously. Your application can be developed inside a sandbox environment and this provides a better option for cases where you don't want to install or modify your host OS for application development.

Using the Remote – WSL extension

To connect to the Linux system, we need to use the Remote – WSL extension. Since we have already installed the Remote extension pack earlier in this chapter, we don't need to install this extension explicitly.

To start with, we need to first enable the WSL on our local machine.

1. Here, we have used Windows 10 as the host OS. The WSL can be enabled from the Windows Features options:

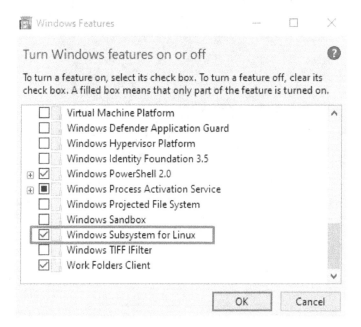

Figure 10.22 – Installing the Windows Subsystem for Linux feature

2. Select **OK** and restart your system. Once this is installed, verify the installation by opening Command Prompt and typing `wsl`. It will show a shell terminal where you can execute Linux commands.

3. Next, we need to install a Linux distro. Go to the Microsoft Store on the Windows machine and install Ubuntu. Once Ubuntu is installed, click on the **Launch** button to start the Ubuntu OS:

Figure 10.23 – Using the Microsoft Store to launch Ubuntu Terminal

You will see a command prompt and it will ask you to enter a username and password to complete the configuration.

Once the configuration is completed, we will now install .NET Core on the Ubuntu OS.

Installing and configuring .NET Core on Ubuntu

We will install .NET Core using the **Advanced Package Tool** (**APT**) in the Ubuntu OS. The APT is a tool used to install software on Ubuntu, Debian, and Linux distributions.

4. To start with, run the following command to add the Microsoft package signing key to the list of trusted keys and add the package repository:

```
wget https://packages.microsoft.com/config/ubuntu/20.04/
packages-microsoft-prod.deb -O packages-microsoft-prod.
deb
sudo dpkg -i packages-microsoft-prod.deb
```

5. Install the SDK by running the following command:

```
sudo apt-get update; \
  sudo apt-get install -y apt-transport-https && \
  sudo apt-get update && \
  sudo apt-get install -y dotnet-sdk-3.1
```

6. Verify the installation by running the .NET Core `version` command:

```
dotnet --version
```

7. Next, create a new folder and name it `dotnetapp`:

```
mkdir dotnetapp
```

8. Go inside the folder and open VS Code. Once the VS Code is loaded, it will start connecting to the Ubuntu machine. VS Code will ask you to install the **Remote – WSL** extension if it's not installed. If so, proceed with the installation and reload VS Code:

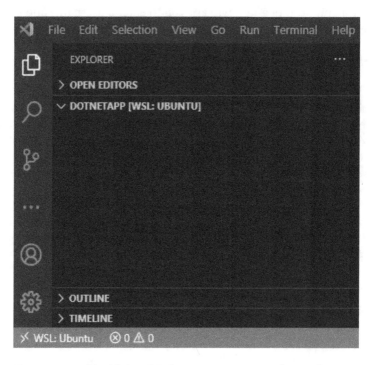

Figure 10.24 – The Ubuntu Workspace once connected using the WSL

Once it is connected, you can create a new .NET Core application using the dotnet CLI.

Open a terminal window from VS Code and type the following command to create a new .NET Core console application project:

```
dotnet new console
```

The preceding command scaffolds a sample .NET Core console application project. Build and run the application and it will all be executed on the Ubuntu machine.

Lastly, once you are done, you can close the remote connection to the Ubuntu machine by selecting **Close Remote Connection** from the command palette:

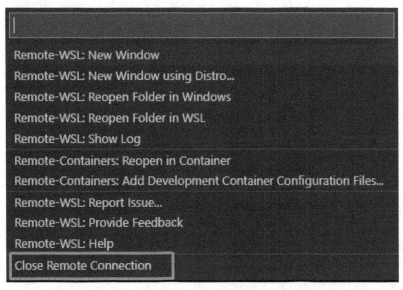

Figure 10.25 – The option to close remote connection from the WSL once done

The preceding command closes the Ubuntu folder and disconnects VS Code from Ubuntu.

In this section, we have learned how easy it is for developers to perform remote development using the SSH, Containers, and WSL extensions in VS Code.

In the next section, we will explore GitHub Codespaces, which allows you to perform remote development in the cloud.

GitHub Codespaces

GitHub Codespaces is designed to provide cloud-based development environments to help develop applications. You need to register on GitHub to be able to use GitHub Codespaces, and then you can develop applications using VS Code.

To get started with GitHub Codespacee, we first need to install the Visual Studio Codespaces extension from the **Extensions** bar in VS Code, shown here:

Figure 10.26 – Installing the Visual Studio Codespaces extension

Once this extension is installed, go to the settings for the Visual Studio Codespaces extension and change the **Codespaces** account provider to **GitHub**:

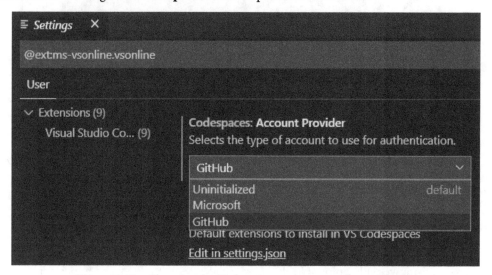

Figure 10.27 – Changing the Codespaces account provider to GitHub

After selecting the account provider, select the **Remote Explorer** icon, , from the sidebar in VS Code, and then sign in to Codespaces. It will prompt you to enter your GitHub credentials and ask you to authorize VS Code to access your GitHub account. You can now create a new GitHub codespace to start developing applications in theCcloud.

At the time of writing this chapter, the GitHub account provider was in preview and the Visual Studio Codespaces extension was in the process of merging with GitHub Codespaces.

In this section, we have covered remote development with GitHub Codespaces using Microsoft as the codespace account provider.

To select GitHub Codespaces with **Microsoft** as the**AaccountPprovider**, you have tgoreturn to the Visual Studio Codespaces extension's settings and select **Microsoft**, as follows:

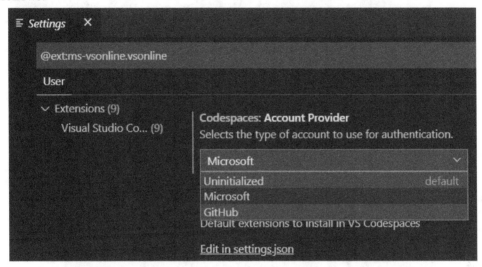

Figure 10.28 – Changing the Codespaces account provider to Microsoft

We can create a new codespace as follows:

Figure 10.29 – Signing in to Codespaces

This will open up the browser and ask you to sign in. Once you are signed in, it will prompt you to open VS Code and it will show the **Signed in to Visual Studio Codespaes.** message, as shown in the following screenshot:

Figure 10.30 – The dialog displayed when successfully signed in to Codespaces

Following the wizard prompt, choose the respective environment to set up your Codespace. The default environment is Linux-based with a 4-core CPU and 8 GB RAM. However, you can also create other codespaces by going through the custom settings. In our case, we will proceed with the **Default settings** option:

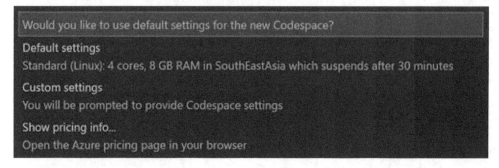

Figure 10.31 – Selecting the Default setting to proceed with Linux OS

After selecting the **Default** settings option, it will ask you to select your Azure subscription, since the environment you create will be created under an active Azure subscription.

Next, we can choose the repository. If you already have an existing repository, you can provide a URL or just click *Enter* to proceed with an empty VS Codespace repository.

Finally, it asks you to enter the name of the repository. Here, let's specify VSCodeBook:

Figure 10.32 – Entering the name of our Codespace as VSCodeBook

After completing the wizard prompts, it starts setting up the repository and you will see the following screen while your Codespace is set up:

Figure 10.33 – Initialization pane of GitHub Codes

Once it is done, it will show a **Connect** button to connect to the Codespace. Alternatively, you can also connect to GitHub Codespaces directly from VS Code:

Figure 10.34 – Workspace in GitHub Codespaces

If you open a VS Code terminal, you will notice the terminal is connected to your Codespace. Since we choose Linux as a default GitHub Codespace, we can run bash commands:

Figure 10.35 – The Terminal window in VS Code showing the Codespace drive

Since Codespace has been set up, we can create a simple .NET Core console application by running the dotnet new command from the Terminal window in VS Code:

```
dotnet new console
```

The preceding command creates a new .NET Core console application project and restores all the dependencies in GitHub Codespaces:

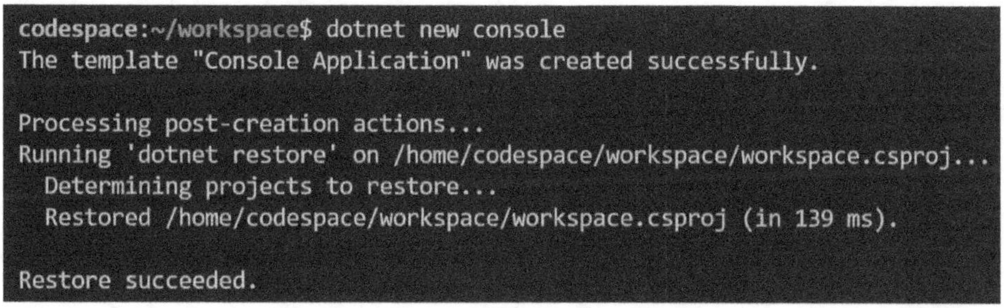

```
codespace:~/workspace$ dotnet new console
The template "Console Application" was created successfully.

Processing post-creation actions...
Running 'dotnet restore' on /home/codespace/workspace/workspace.csproj...
  Determining projects to restore...
  Restored /home/codespace/workspace/workspace.csproj (in 139 ms).

Restore succeeded.
```

Figure 10.36 – The .NET Core Console App is successfully created and restored all the dependencies

Finally, run the `dotnet` application using the following command:

```
dotnet run
```

This command will run the .NET Core console application created in the preceding steps and displays a **Hello World!** message:

```
codespace:~/workspace$ dotnet run
Hello World!
codespace:~/workspace$ []
```

Figure 10.37 – The .NET Core Console Application running successfully on GitHub Codespaces

If you access the Explorer in VS Code, you will notice the files are there and you can add more or customize them to build the application:

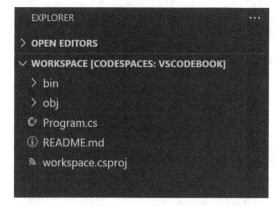

Figure 10.38 – The VS Codespace workspace and .NET Core Console App files

Lastly, you can also access your Codespaces by opening the **REMOTE EXPLORER** tab and selecting **Codespaces** from the dropdown, and can then further connect to your Codespace (if not already connected):

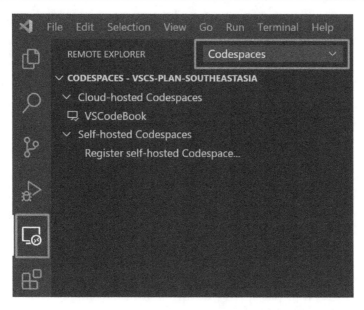

Figure 10.39 – Access your Codespaces by selecting the Remote Explorer tab and choosing Codespaces from the dropdown

> **Tip**
> If you want to learn more about GitHub Codespaces, check the following link:
> ```
> https://code.visualstudio.com/docs/remote/
> codespaces
> ```

Whether you are working on a long- or short-term project, you can use VS Codespace to create environments in the cloud and develop applications without installing any development frameworks, dependencies, and so on locally on your machine.

Summary

In this chapter, we learned about remote development and how this is possible in VS Code using the remote development extension pack. We understood various scenarios in which we could do remote development on a remote machine, a container, or even in the cloud by using GitHub Codespaces.

We used the Remote – SSH extension to develop a .NET Core application remotely on macOS; used the Remote – Containers extensions to develop a Node.js application inside an isolated, sandbox containerized environment; and used the Remote – WSL extension to develop a .NET Core application in the WSL.

Lastly, we learned about GitHub Codespaces and set up a new Codespace from VS Code on Linux and developed a simple .NET Core console application.

With this, we have ended our journey of developing apps with VS Code. We hope this book has helped you delve deeper into the applications we've shown you, and that you want to try them and develop more!

Other Books You May Enjoy

If you enjoyed this book, you may be interested in these other books by Packt:

Hands-On Software Architecture with C# 8 and .NET Core 3

Gabriel Baptista, Francesco Abbruzzese

ISBN: 978-1-78980-093-7

- Overcome real-world architectural challenges and solve design consideration issues
- Apply architectural approaches like Layered Architecture, service-oriented architecture (SOA), and microservices
- Learn to use tools like containers, Docker, and Kubernetes to manage microservices
- Get up to speed with Azure Cosmos DB for delivering multi-continental solutions
- Learn how to program and maintain Azure Functions using C#
- Understand when to use test-driven development (TDD) as an approach for software development
- Write automated functional test cases for your projects

ASP.NET Core 3 and Angular 9 – Third Edition

Valerio De Sanctis

ISBN: 978-1-78961-216-5

- Implement a Web API interface with ASP.NET Core and consume it with Angular using RxJS Observables

- Create a data model using Entity Framework Core with code-first approach and migrations support

- Set up and configure a SQL database server using a local instance or a cloud data store on Azure

- Perform C# and JavaScript debugging using Visual Studio 2019

- Create TDD and BDD unit test using xUnit, Jasmine, and Karma

- Implement authentication and authorization using ASP.NET Identity, IdentityServer4, and Angular API

- Build Progressive Web Apps and explore Service Workers

Leave a review - let other readers know what you think

Please share your thoughts on this book with others by leaving a review on the site that you bought it from. If you purchased the book from Amazon, please leave us an honest review on this book's Amazon page. This is vital so that other potential readers can see and use your unbiased opinion to make purchasing decisions, we can understand what our customers think about our products, and our authors can see your feedback on the title that they have worked with Packt to create. It will only take a few minutes of your time, but is valuable to other potential customers, our authors, and Packt. Thank you!

Index

www.ingramcontent.com/pod-product-compliance
Lightning Source LLC
LaVergne TN
LVHW060039070326
832903LV00072B/1137